THE ARTILECT WAR

Cosmists vs. Terrans

A Bitter Controversy Concerning
Whether Humanity Should Build
Godlike Massively Intelligent Machines

Hugo de GARIS, Ph.D.

An **etc** *Publication*

Library of Congress Cataloging-in-Publication Data

De Garis, Hugo, 1947-
    The artilect war : cosmists vs. terrans : a bitter controversy
concerning whether humanity should build godlike massively
intelligent machines / Hugo de Garis.
        p. cm.
Includes bibliographical references.
ISBN 0-88280-153-8 -- ISBN 0-88280-154-6
1.   Science--Moral and ethical aspects. 2. Artificial intelligence.
I. Tttle. 0180.55.M67D4 2005

The Author's Websiite and Email Contact Address

Prof, Hugo de GARIS, Ph.D.

http:/www.cs.usu.edu/-degaris

theartilectwar@yahoo.com

Published by ETC Publications
            Palm Springs
            California 92262

The front cover shows Professor de Garis with his first
generation Brain Building Machine—Cam-Brain Machine
(CBM).

Published in the United States of America.

CONTENTS

FOREWORD

Hugo de Garis and I go back a long way. We have often appeared together over the years as invited speakers at international gatherings of futuristic thinkers. What we have in common is the belief that later this century, humanity will have to confront the prospect of being replaced by a new dominant species, namely, ultra intelligent robots controlled by ultra intelligent artificial brains. Where Hugo and I differ is that (using his terminology) he is primarily a "Cosmist" (someone who believes, in his words, that "godlike massively intelligent machines," should be built, no matter what the risk to humanity's future survival). Hugo would, I think, label me as a "Terran" (someone who is opposed to the idea that Cosmists should be allowed to build such ultra intelligent machines).

I remember a conversation with him at a recent conference we both happened to be at. He asked me whether I was a Cosmist or a Terran. I said that Terran was nearer the mark. Hugo then bristled and said half jokingly, "I guess that's how it starts!" He was referring to his "Artilect War" scenario that this book is about. He felt that a major war will brew between the Cosmists and the Terrans late this century over whether humanity should or should not build these godlike machines that he calls "artilects" (artificial intellects).

In actual fact, I would consider myself more as a "Cyborgian" than a "Terran." ("Cyborgians" are people who look to technically upgrade their bodies to become "cyborgs," i.e. part machine, part human.). I hope that by enhancing ourselves, we humans can have our cake and eat it too by achieving the dream of attaining the godlike abilities that Hugo talks about by converting ourselves into part-artilects, without having to pay the cost of a major war. In a sense I am looking at a sort of compromise – rather than having ultra intelligent artificial brains acting against humanity, we join with them.

A few years ago, Hugo and I gave invited talks at a venue in Switzerland. Half jokingly, and to make a point, Hugo had arranged for the organizers of the meeting to supply him with a child's cowboy gun, which he then used to "shoot" me, once we had finished our talks. A photo of this event can be found on Hugo's website (at http://www.cs.usu.edu/~degaris/news/zurich.html). It is entitled "The First Shot in the Artilect War." I sincerely hope it will not come to this sort of end game in the real world. Hugo's scenario of a major war late this century, in which literally billions of people die, due to the use of advanced 21st century weaponry, is extremely depressing, and I firmly hope he is wrong, dead wrong, for the sake of humanity's (and cyborgian) survival.

After the reader has finished studying this book, serious doubts may well arise as to whether my more optimistic scenario is more likely to win out. Hugo's reasoning is frighteningly persuasive, even though my viscera reject what he is saying. The fact that he is pioneering the new field of "artificial brains" only increases the credibility of his vision. If anyone in the world is in a good position to predict the future impact of advanced artificial brains on humanity, it must be Hugo.

I believe that this book is of profound importance. If many decades into the future, Hugo is proven to be correct in saying that "the species dominance issue will dominate our global politics this century," then he will have become one of the major thinkers of the 21st century. With no offence to Hugo, I hope that he will be shown to have been wrong, to be shown to have exaggerated, and overreacted; because if not, the fate that he is suggesting will befall our grandchildren, is too horrible to grasp fully for all humanity, what he would call "gigadeath!"

No matter where you sit in the Cosmist-Terran opinion spectrum, this is a book that cannot and should not be ignored. It is too important and too disturbing to be summarily dismissed. I advise that you read it, then read it again, and try to find faults with Hugo's logic and judgment, so that we can all look forward to a peaceful prosperous 21st century. Because, if Hugo's vision is correct, the hellish nightmare, as portrayed in films such as The Terminator, will become a reality.

Kevin Warwick, Ph.D.
Professor of Cybernetics, Reading University, England.
Author of   *"I, Cyborg"*, *"In the Mind of the Machine"*,
"March of the Machines"

Chapter 1

INTRODUCTION

My name is Professor Hugo de Garis. I'm the head of a research group which designs and builds "artificial brains," a field that I have largely pioneered. But I'm more than just a researcher and scientist -- I'm also a social critic with a political and ethical conscience. I am very worried that in the second half of our new century, the consequences of the kind of work that I do may have such a negative impact upon humanity that I truly fear for the future.

You may ask, "Well, if you are so concerned about the negative impact of your work on humanity, why don't you just stop it and do something else?" The truth is, I feel that I'm constructing something that may become rather godlike in future decades (although I probably won't live to see it). The prospect of building godlike creatures fills me with a sense of religious awe that goes to the very depth of my soul and motivates me powerfully to continue, despite the possible horrible negative consequences.

I feel quite "schizophrenic" about this. On the one hand I really want to build these artificial brains and to make them as smart as they can be. I see this as a magnificent goal for humanity to pursue, and I will be discussing this at length in this book. On the other hand, I am terrified at how bleak are some of the scenarios that may ensue if brain building becomes "too successful," meaning that the artificial brains end up becoming a

1

lot more intelligent than the biological brains we carry around in our skulls. I will be discussing this too at length in this book.

Let me be more specific. As a professional brain building researcher and former theoretical physicist, I feel I am in a position to see more clearly than most the potential of 21st century technologies to generate "massively intelligent" machines. By "massively intelligent" I mean the creation of artificial brains which may end up being smarter than human brains by not just a factor of two or even ten times, but by a factor of trillions of trillions of trillions of times, i.e. truly godlike. Since such gargantuan numbers may sound more science fiction like to you than any possible future science, the third chapter of this book will explain the basic principles of those 21st century technologies that I believe will allow humanity, if it chooses, to build these godlike machines. I will try to persuade you that it is not science fiction, and that strong reasons exist to compel humanity to believe in these astronomically large numbers. I will present these technologies in as simple and as clear a way as I can, so that you do not need to be a "rocket scientist" (as the Americans say, i.e. someone very smart) to understand them. The basic ideas can be understood by almost anyone who is prepared to give this study a little effort.

The third chapter introduces you to all these fabulous 21st century technologies that will permit the building of godlike massively intelligent machines. Probably a host of ethical, philosophical, and political questions will occur to you. The prospect of humanity building these godlike machines raises vast and hugely important questions. The majority of this book is devoted to the discussion of such questions. I don't pretend to have all the answers, but I will do my best.

2

One of the great technological economic trends of our recent history has been that of "Moore's law," which states that the computational capacities (e.g. electronic component densities, electronic signal processing speeds, etc) of integrated circuits or "chips," have been doubling every year or two. This trend has remained valid since Gordon Moore, one of the founders of the Intel microprocessor manufacturing company, first formulated it in 1965. If you keep multiplying a number by 2 many times over, you will soon end up with a huge number. For example, 2 times 2 times 2 times 2 … (ten times) equals 1024. If you do it 20 times you get 1048576, i.e. over a million. If you do it 30 times, you get a billion, by 40 times you get a trillion, etc. Moore's law has remained valid for the past four decades, so that the size of the doublings recently has become truly massive. I speak of "massive Moore doublings."

Moore's law is a consequence of the shrinking of the size of electronic circuits on chips, so that the distance that electrons (the elementary particles whose flow in an electronic circuit is what constitutes the electrical current) have to travel between two electronic components, (for example two transistors), is reduced. According to Einstein, the fastest speed at which anything can move is the speed of light (about 300,000 kms/sec) and this is a constant of nature that electronic currents have to respect. If one shortens the distance between two electronic components, then an electronic signal between them (i.e. the flow of electrons between them) has less distance to travel, and hence takes less time to traverse that distance (at the constant speed of light).

A huge effort over the past few decades has been devoted by the chip manufacturing companies into making electronic circuits smaller, and hence denser, so that they function faster. The faster a microprocessor chip functions, the more economically attractive it is. If you are the CEO of a chip manufacturing company and your competitor down the road in California's "Silicon Valley" brings a rival chip onto the market that is 30% faster than yours and six months ahead of you, then your company will probably go out of business. The market share of the rival company will increase significantly, because everyone wants a faster computer. Hence, for decades, electronic circuitry has become smaller and thus faster.

How much longer can Moore's law remain valid? If it does so until 2020, then the size of the electronic components in mass memory chips will be such that it will be possible to store *a single bit of information* on a *single atom.* (A "bit" is a "binary digit," a 0 or a 1, that computers use to represent numbers and symbols to perform their calculations.) So how many atoms (and hence how many stored bits) are there in a human sized object, such as an apple? The answer is astonishing -- a trillion trillion atoms (bits), i.e. a 1 followed by 24 zeros, or a million million million million.

Are you beginning to get an inkling for why I believe that massively intelligent machines could become trillions of trillions of times smarter than we are later this century?

Not only is it likely that 21st century technology will be storing a bit of information on a single atom, it will be using a new kind of computing called "quantum computing," which is radically different from the garden variety or "classical computing" that humanity used in the 20th century. The third chapter will attempt to give a brief outline of the principles of quantum computing since it is likely that that technology will form the basis of the computers of the near and longer term future.

The essential feature of quantum computing can however be mentioned here. It is as follows. If one uses a string of N bits (called a "register" in computer science, e.g. 001011101111010) in some form of computing operation (it doesn't matter for the moment what the operation is) it will take a certain amount of time using "classical computing." However in the same amount of time, using "quantum computing" techniques, one can often perform 2^N such operations. (2^N means 2 multiplied by 2 multiplied by 2 ... (N times)). As N becomes large, 2^N becomes astronomically large. The potential of quantum computing is thus hugely superior to classical computing. Since Moore's law is likely to take us to the atomic scale, where the laws of physics called "quantum mechanics" need to be applied, humanity will be forced to compute quantum mechanically, hence the enormous theoretical and experimental effort in the past few years to understand and build "quantum computers."

Quantum computing still has many conceptual and practical problems that need to be solved before quantum computers are sold to the public. But progress is being made every month, so personally I believe that it is only a question of time before we have functional quantum computers.

Now, start putting one bit per atom memory storage capacities together with quantum computing and the combination is truly explosive. 21st century computers could have potential computing capacities truly trillions of trillions of trillions .... of times above those of current classical computing capacities.

I hope you have followed me so far.

At this point in the argument, you may be racing ahead of me a little and object that I seem to be assuming implicitly that massive memory capacities and astronomical computational capacities are sufficient to generate massively intelligent machines, and that nothing else is needed. I have been accused by some of my colleagues of this, so let me state my personal opinion on this question.

There are people (for example, Sir Roger Penrose, of black hole theory fame, and arch rival of the British cosmologist Stephen Hawking) who claim that there is more to producing an intelligent machine than just massive computational abilities. Penrose claims that consciousness would also be needed, and that new physics will be required to understand the nature and creation of artificial consciousness in machines.

I am open to this objection. Perhaps such critics are right. If so, then their objections do not change my basic thesis very much, perhaps causing a delay of several decades as the nature of consciousness is better understood. I feel that it is only a question of time before science understands how nature builds us, i.e. I expect science will come to understand the "embryogenic" process, used in building an embryo and then a baby, consisting of trillions of cells, from a single fertilized egg cell.

We have the existence proof of ourselves, who are both intelligent and conscious, that it is possible for nature to assemble molecules in an appropriate way to build us. When a pregnant woman eats, some of the molecules in her food are rearranged, and then self assembled into a large molecular structure consisting of trillions of trillions of atoms that becomes her baby. The baby is a self-assembled collection of molecules that gets built to become a functional three-dimensional creature that is intelligent and conscious.

Nature, i.e. evolution, has found a way to do this, therefore it can be done. If science wants to build an intelligent conscious machine, then one obvious strategy is to copy nature's approach as

6

closely as possible. Sooner or later, science will end up with an artificial life form that functions in the same way as a human being.

Common sense says that it would be easier to build an artificial brain, if science had a far better knowledge of how our own biological brains work. Unfortunately, contemporary neuroscience's understanding of how our brains work is still painfully inadequate. Despite huge efforts of neuroscientists over the past century or more to understand the basic principles of the functioning of the human brain, very little is known at the neural micro-circuit level as to just how a highly interconnected neural circuit does what it does. Science does not yet have the tools to adequately explore such structures.

However, as technology becomes capable of building smaller and smaller devices (moving down from the micro-meter level to the nano-meter level (i.e. from a millionth of a meter (the size of bacteria) to a billionth of a meter (the size of molecules)) it will become possible to build molecular scale robots that can be used to explore how the brain functions.

Science's knowledge of how the biological brain works is inadequate because the tools we have at our disposal today are inadequate, but with molecular scale tools (called "nanotech" or "nanotechnology") neuroscientists will have a powerful new set of techniques with which to explore the brain. Progress in our understanding of how the brain functions should then be rapid.

Brain builders like me will then jump on such newly established neuro-scientific principles and incorporate them rapidly into our artificial brain architectures.

Hopefully in time, so much will become known about how our own brains function, that a kind of "intelligence theory" will arise, which will be able to explain on the basis of neuronal circuitry (a neuron is a brain cell) why Einstein's brain for example, was so much smarter than most other people's brains. Once such an intelligence theory exists, it may be possible for neuro-engineers like myself to take a more engineering approach to brain building. We will not have to remain such "slaves to neuroscience." We will be able to take an alternative route to producing intelligent machines (although admittedly initially based on neuro-scientific principles).

So with the new neuro-scientific knowledge that nanotech tools will provide, and the computational miracles that quantum computing and one bit per atom storage will allow, brain builders like me will probably have all the ingredients we need to start building truly intelligent and conscious machines.

At this point, a host of questions arises, and I will spend most of this book trying to answer a lot of them. Lets jump into the future for a moment and try to imagine how the above technological developments will impact on ordinary peoples lives.

Pretty soon, it will be possible to buy artificially brained robots that perform useful tasks around the house. If the price of such robots can be made affordable, then the demand for them will be huge. I believe in time that the world economy will be based upon brain-based computers. Such devices will be so useful and so popular that everyone on the planet will want to own them. As the technologies and the economics improve, the global market for such devices will only increase to the point that most of the planet's politics will be tied up in supporting it. Not only will the commercial sector be heavily involved in the production of ever smarter and ever more useful robots and artificial brain based devices, but so too of course will the military forces of the world.

It is unlikely in the next few decades that the planet will have formed a truly global state, with a global police force to defend its global laws. Instead I believe there will be a growing political rivalry over the next half century between the United States and China to be the world's most powerful nation. This rivalry will ensure that the ministers of defense of both countries cannot afford to allow the other country to develop more intelligent soldier robots and other artificial brain based defense systems than their own. Hence, national governments will be heavily involved in pushing the development of military based artificial brain research that will only spill over in time to the commercial sector, as has been the pattern for over a century.

Thus the rise of artificial brain based robotics and related fields, seems unstoppable. There will be so much military and commercial momentum behind it that it is difficult to imagine how it could be stopped, unless somehow a mass political movement is formed to block its development.

How might such a movement get off the ground? It's not too difficult to imagine what might happen. Imagine a few decades from now that millions of people have already bought household cleaning robots, sex robots, teaching machines, babysitter robots, companionship robots, friendship robots, etc, and that these brain based machines talk quite well and understand human speech to a reasonable extent. A few years later what happens? Not surprisingly, the models of that earlier year are now seen by their owners to be rather old fashioned and not as attractive as the latest models. The latest models will be more "intelligent" because their speech is of higher quality. They will understand more and give better, more appropriate answers. Their behavioral repertoire will be richer. In short, they will make the earlier models look quite inferior.

So what does everyone do? Of course, they will scrap their old robots and buy new ones, or have their old ones updated with better artificial neural circuitry. In a further few years, the same process will repeat itself, in a fashion similar to the way buyers of personal computers behaved in the 1980s and 1990s, etc.

However, some of the more reflective buyers may start noticing that their household machines and robots are becoming smarter and smarter in every machine generation so that the IQ gap between human beings and robots keeps getting smaller. Once the robots start getting really quite smart, suddenly millions of robot owners will start asking themselves some awkward questions.

"Just how smart could these artificially brained robots become?"

"Could they become as smart as human beings?"

"If that's possible, is that a good thing?"

"Might not the robots then be smart enough to be a threat to humanity?"

"Could the robots become smarter than humans?"

"If so, how much smarter?"

"Should humanity allow these robots to become smarter than human beings?"

"If they become a lot smarter than human beings, might they decide that humans are a pest, a cancer on the surface of the planet, and decide to wipe us out?"

"Should humanity take the risk, that that might happen?"

"Should a limit be placed on the robot's AIQ (Artificial Intelligence Quotient), so that the robots are smart enough to be useful to human beings, but not too smart so as to be threatening?"

"Will it be possible to stop the rise of robot AIQ?"

10

"Will it be politically, militarily, economically possible to stop the robots becoming smarter every year?"

"There are lots of people who see the creation of massively intelligent machines as the destiny of the human species. These people will not like any limits being placed on AIQ levels. Won't this create conflict amongst human beings?"

You may be able to think of other such questions relating to the rise of artificial intelligence and the creation of artificial brains with ever-greater capabilities.

How do I see humanity facing up to the challenge of the rise of smart machines? My personal scenario that I find the most plausible I will present to you now. However, before doing so, I would like to introduce a new term that I will use from now on throughout this book, as it is a useful shorthand for the term "godlike massively intelligent machine." The new term is "artilect," which is a shortened version of "artificial intellect." The term "artilect" features in the very title of this book "The Artilect War," so it is probably the most important concept and term in this book.

I believe that the 21st century will be dominated by the question as to whether humanity should or should not build artilects, i.e. machines of godlike intelligence, trillions of trillions of times above the human level. I see humanity splitting into two major political groups, which in time will become increasingly bitterly opposed, as the artilect issue becomes more real and less science fiction like.

The human group in favor of building artilects, I label the "Cosmists," based on the word "cosmos" (the universe), which reflects their perspective on the question. To the Cosmists, building artilects will be like a religion; the destiny of the human species; something truly magnificent and worthy of worship; something to dedicate one's life and energy to help achieve. To the Cosmists, not building the artilects, not creating the next higher form of evolution, thus freezing the state of evolution at the *puny* human level, would be a *"cosmic tragedy."* The Cosmists will be bitterly opposed to any attempt to stop the rise of the 21st century artilect.

The second human group, opposed to the building of artilects, I label the "Terrans," based on the word "terra" (the Earth) which reflects their inward looking, non-cosmic, perspective. The Terrans, I strongly suspect, will argue that allowing the Cosmists to build their artilects (in a highly advanced form) implies accepting the *risk*, that one day, the artilects might decide, for whatever reason, that the human species is a pest. Since the artilects would be so vastly superior to human beings in intelligence, it would be easy for the artilects to exterminate the human species if they so decided.

But you may argue that if the artilects truly become very smart, they would realize that human beings gave birth to them, that we are their parents. Therefore the artilects would respect us and treat us well. This may be what happens, but the point is, I argue, that you could not be certain that the artilects would treat humanity with the level of respect that we would like.

Don't forget, the artilects have the potential of becoming trillions of trillions .... of times smarter than we are, so there is always the possibility that they could become so smart that human beings would appear to them to be so inferior that we would simply not be worth worrying about. Whether humanity survives or not, might be a matter of supreme indifference to them.

It is not exaggerating to say that there is quite a close analogy between an artilect trying to communicate with a human being, and a human being trying to communicate with a rock.

To make another analogy, consider your feelings towards a mosquito as it lands on the skin of your forearm. When you swat it, do you stop to consider that the creature you just killed is a miracle of nano-technological engineering, that scientists of the 20th century had absolutely no way of building. The mosquito consists of billions of cells, each of which can be looked upon as a kind of molecular city, where a molecule in a cell is equivalent to a person in a city. The comparative scale of molecule to cell is about the same as person to city.

Despite the fact that the mosquitoes, which took billions of years to evolve, are extremely complex and miraculous creatures, we human beings don't give a damn about them, and swat them because from our perspective they are a pest. We have similar attitudes towards killing ants when we walk on them during a stroll through the forest, or when flushing spiders down the plughole.

Who is to say that the artilects might not have similar attitudes towards human beings, and then wipe us out? With their gargantuan "artilectual" intelligence, it would be as easy as pie for them to do so.

The critical word in the artilect debate for the Terrans is "risk." The Terrans will argue that humanity should never take the risk that the artilects, in an advanced form, might decide to wipe out the human species. The only certain way that the risk remains zero is that the artilects are never built in the first place.

When push comes to shove, if the Terrans see that the Cosmists are truly serious about building artilects in an advanced state, then to preserve the survival of the human species, the Terrans will exterminate the Cosmists. Killing a few million

Cosmists will be considered justifiable by the Terrans for the sake of preserving the survival of the whole human species, i.e. billions of people.

Such a sacrifice would be deemed reasonable by the Terrans. To make a historical analogy -- when Stalin's troops were pushing west at the end of WWII, to capture Berlin and destroy Hitler's Nazi regime that murdered 20 million Russians, they were losing about 100,000 Russian soldiers killed or injured for every major east European city captured from the Nazis. To Stalin, such a sacrifice was considered justifiable for the greater good of ridding the Russian people of the horror of mass murdering Nazism.

You may now ask, "Would anyone in their right mind genuinely choose, when push comes to shove, to be a Cosmist, and truly risk the annihilation of the human species?"

I think that in the future, millions of people will answer yes to this most fundamental of questions. I think that as more people become fully conscious of what the artilects could become, many of these people will end up choosing in favor of their creation. This book will devote a whole chapter to arguments in favor of building artilects when it presents the Cosmist case.

These people, these "Cosmists," will place a higher priority on the creation of godlike, immortal, go anywhere, do anything creatures (where one artilect is "worth" a trillion trillion human beings) than running the risk of seeing the extermination of the human species at the hands of the artilects.

Let me spell this out, so that there is no doubt about the stance of the Cosmists. A Cosmist, by definition, is someone who favors the building of artilects. The artilects, if they are built, may later find humans so inferior and such a pest, that they may decide, for whatever reason, to wipe us out. Therefore *the Cosmist is prepared*

14

to accept the risk that the human species is wiped out. If humanity is wiped out, that means your grandchildren will be wiped out, my grandchildren will be wiped out. It would be the worst calamity in human history, because there would be no more history, because there would be no more humans. Humanity would thus join the long list of over 99% of species that have ever existed on the Earth, that have already become extinct.

Thus to the Terrans, the Cosmists are monsters incarnate, far worse than the regimes of Hitler, Stalin, Mao, the Japanese, or any other regime that murdered tens of millions of people in the 20th century, because the scale of the monstrosity would be far larger. This time we are not talking about deca-mega mass murder, we are talking about the potential annihilation of the whole human species, billions of people.

But to the Cosmists, the survival or not of the human species, on an insignificant planet, circling a star that is one of about 200 billion in our galaxy, in a known universe of a comparable number of galaxies (also in the billions), and with probably as many universes in the "multiverse" (according to several recent cosmological theories) is a matter of miniscule importance. I have labeled the Cosmists "Cosmists" for a reason. Their perspective is cosmic. They will look at the "big picture" -- meaning that the annihilation of one ultra-primitive, biological, non-artilectual species (i.e. human beings) on one insignificant little planet, is unimportant in comparison with the creation of artilect gods.

Such ideas and attitudes will be elaborated upon in two chapters later in this book, presenting the Terran and the Cosmist cases, one for each viewpoint. There are very powerful arguments on both sides, which I believe will only make the inevitable conflict between Terranism and Cosmism all the more bitter as the

15

artilect debate heats up in the coming decades.

What makes me particularly gloomy about the potential bitterness of this coming conflict is how evenly people's opinions are split along the Terran/Cosmist divide. For example, I often invite audiences to whom I present the Cosmist/Terran/Artilect scenario in public lectures, to vote on whether they would be Terran or Cosmist. I find that the voting is not what I first expected it would be (namely about 10% Cosmist, 90% Terran) but rather 50/50, 60/40, 40/60. This issue truly divides people.

What makes me even gloomier is that the artilect issue (i.e. should artilects be built or not) will heat up in the 21st century to such an extent, that it is almost certain it will lead to a major war between the Terrans and the Cosmists in the second half of this new century. *This conflict will take place with 21st century weaponry.* If one extrapolates up the graph of the number of deaths in major wars from the beginning of the 19th century (e.g. the Napoleonic wars) to the end of the 21st century, one arrives at the depressing figure of billions, what I call "gigadeath."

But the population of the Earth is only several billion people, so we arrive at the tragic conclusion that to avoid the risk of the total annihilation of the human species by the artilects, humanity goes to war against itself and kills itself off (or almost).

This "Artilect War" as I call it, will be the most passionate in history, because the stake has never been so high, namely the survival of the whole human race. It will be waged with 21st century weapons and hence the casualty figures will be of 21st century grandeur.

The sad thing about this gloomy scenario is that despite considerable effort on my part, I have been unable to find a way out of this mess. I lie awake at night trying to find a realistic

16

scenario that could avoid "gigadeath." I have not succeeded, which makes me feel most pessimistic. In fact I am so pessimistic that I am glad to be alive today. At least I will die peacefully in my bed. However I fear for my grandchildren. They may well see the horror of it and very probably they will be destroyed by it.

I will die within about 30-40 years, given my age, but that is not enough time I believe, for the artilect scenario to unfold. I believe it will take longer than that to obtain the necessary knowledge to build massively intelligent artificial brains or artilects. However, what I will see in my lifetime, and obviously this book is aimed at producing just that, is a vociferous debate over the artilect issue.

There are a growing number of researchers and professors like myself who are starting to see the writing on the wall, and who are claiming publicly in media appearances and books that the 21st century will see the rise of massive artificial intelligence. I am the only one so far who is saying that this rise of massive AI will probably lead to a major war, the "Artilect War."

Thus the issue is really starting to hit the world media, and countries such as the US, the UK, France and Holland are leading the pack. In fact I believe that within only a few years, the issue will have passed from one that is confined largely to academic audiences, to a wider general public, with contributors from such fields as politics, religion, defense, etc.

The "Artilect War" will seem like science fiction, and seem to be set too far into the future, for most people to worry about, but as the machines start getting smarter and smarter every year, it will take on an intensity that will become truly frightening.

So what is my position on all this? Why am I writing this book? Deep down, I'm a Cosmist. I think it would be a cosmic tragedy if humanity chooses never to build artilects. To illustrate my reasons for being a Cosmist, I like to tell a little story.

Imagine you are an ET (an extra terrestrial) with godlike technological powers, and you come to the Earth three billion years ago. You observe the life forms at that time on Earth and notice that they are still at the primitive bacterial single-celled stage. In a sweep of your magical technological wand, you fiddle all the DNA in all the bacteria of the planet so that (for the sake of the argument) it will never be possible for these bacteria to evolve into multi-celled creatures in the future. Hence, there will never be any plants, no animals, no human beings, no Einstein, no Beethoven's 9th. Is that a tragedy? Once the multi-celled creatures did evolve on the Earth, zillions of bacteria were eaten by them. The evolutionary rise of multi-celled creatures on the Earth was no picnic for the bacteria.

I hope you see the analogy. If we build artilects and billions of human beings are wiped out as a result, what will be the equivalent of Beethoven's 9th that the artilects will produce with their godlike intellects? As human beings, we are too dumb to know. We are just too inferior to be capable of recognizing such things. It would be like asking a mouse to study Einstein's General Theory of Relativity. It just couldn't do it, because it doesn't have the necessary neural circuitry to allow it to, nor do most humans, for that matter.

But, you may ask, if I'm a Cosmist at heart, why am I writing this book? The answer is that I'm not 100% Cosmist. If I were, I would be quietly getting on with my brain building work and not trying to raise the alarm on the artilect issue to the general public. Part of me is also Terran. On my deathbed I would be proud to be considered the "father of the artificial brain," but if history

18

condemns me as being the "father of gigadeath," then that prospect truly horrifies me. My second wife's mother was gassed by the Nazis at Auschwitz. I know to some extent what genocide means at an emotional level, and have had to live with its consequences for years.

I'm writing this book to raise the alarm, because I think humanity should be given the choice to stop the Cosmists before they get too advanced in their work, *if that is what most human beings choose.* So should I stop my brain building work now? No. I don't think so. I believe that producing near human-level artificial intelligence is a very difficult problem that will take decades to solve. Over the next 30 to 40 years, it is likely that the AIQ of robots will become high enough so that the robots become very useful to humanity. They will perform many of the boring, dirty and dangerous tasks. Humanity will be liberated from such work, and hence have more time to pursue more rewarding tasks. The robots will do most of the work, allowing human beings to do more fun things.

It would be premature to stop the research on artificial brains now. However, once these artificial brains really do start becoming smart and threaten to become a lot smarter and perhaps very quickly (a scenario called "the singularity") then humanity should be ready to take a decision on whether to proceed or not. Making an informed decision on an issue that concerns the survival of the whole human species is something so important that the necessary discussion on the artilect issue should begin early. There should be enough time for all the issue's intricacies to be thrashed out before the artilect age is imminent.

So publicly I'm Terran. I'm trying to raise the alarm. Privately I'm Cosmist. Hence I feel quite schizophrenic, as I mentioned in the very first page of this book. I feel so torn on the issue, so ambivalent. I believe that similar feelings will be felt by billions of people in the future as the artilect debate really takes hold. From the Terran viewpoint, to be a Cosmist is to be a "speciecidal monster" (a species killer). A Cosmist accepts the risk of seeing the human species wiped out by the artilects. This is inherent in the nature of the situation. The decision whether to build artilects has a binary answer – we can build them or not build them. The decision to build them is the decision to accept the *risk* that they may wipe us out.

On the other hand, not to build them is the decision not to build gods, a kind of "deicide" (god killing). From the Cosmist viewpoint, Terrans are "deicidal monsters."

In passing, I should mention that there are some people who feel that the whole Cosmist/Terran conflict can be avoided by having human beings themselves become artilects by adding components to their heads etc to become "cyborgs" (cybernetic organisms, i.e. part human, part machine). Personally I find such arguments naïve, since they would only work if the whole of humanity made the transition from human to artilect at the same rate, which obviously is not going to happen.

There is more potential computing capacity in a grain of sugar than there is in the human brain by a factor of trillions. Incorporating such a grain into the human brain would simply make the human cyborg an "artilect in human disguise" as seen from the perspective of a Terran. The Terrans would hate the cyborgs with as much venom as they would the artilects and would be motivated to destroy both. Having a human exterior would not make the cyborgs any less threatening to the Terrans.

Let me try to express this Terran revulsion against the cyborgs in an even more graphic way that may have a stronger appeal to women than to men. Take the case of a young mother who has just given birth. She decides to convert her baby into a cyborg, by adding the "grain of sugar" to her baby's brain, thus transforming her baby into a human faced artilect. Her "baby" will now spend only about a trillionth of its mental capacity thinking human thoughts, and the rest of its brain capacity (i.e. 99.9999999999% of it) will be used for thinking artilect thoughts (whatever they are). In effect, the mother has "killed" her baby because it is no longer human. It is an "artilect in human disguise" and *totally* alien to her.

Thus to me, the cyborg option will not avoid the Cosmist/Terran conflict. If anything, it will probably only worsen it, because it will increase the level of paranoia of the Terrans when they cannot distinguish easily a cyborg from a human at a distance.

For about ten years I sat on the fence, presenting my ideas in a "on the one hand, on the other hand" kind of way, presenting the two cases, one in favor of the Terrans, and the other in favor of the Cosmists. After some years, my friends began to accuse me of being a hypocrite. "Hugo, you expect humanity to choose between being Terran or Cosmist, but you don't do the same yourself." "Fair enough," I thought, so I chose. In my heart I'm a Cosmist, and I'll try to present the many arguments and feelings in favor of building artilects in the chapter on the Cosmist viewpoint (Ch. 4.) In that chapter I will try to justify why I and other Cosmists feel so passionately about building artilects, that we are prepared to run the terrible risk of the extermination of the human species.

In the chapter on the Terran viewpoint (Ch. 5), I will present the case why the Terrans feel that building artilects would be a total disaster.

21

Later on in this book, I will try to paint a picture as to how I see the conflict brewing and what the possible outcome might be (Ch. 6).

But at this stage I can imagine that some of you may be having a hard time imagining how it might be possible for machines to become trillions of trillions of times smarter and more capable than human beings. It seems so science fiction like, and not to be taken seriously.

This is not the first time this kind of thing has occurred.

In 1933, a Hungarian Jewish physicist by the name of Leo Szilard was staying in a London hotel reading a newspaper report of a talk given by the famous New Zealand physicist, Lord Rutherford, the discoverer of the atomic nucleus. He was asked by a journalist attending the talk if he felt that the day would come when the incredible energy lurking in the nucleus would be tapped at an industrial scale. Rutherford's famous reply was "moonshine", i.e. "no way," "impossible," "rubbish". Szilard felt skeptical about this and felt there had to be a way.

Crossing a London street one day, he had an epiphany. The year was 1933, the year after the neutron was discovered. He realized that the neutron would make a wonderful nuclear bullet that would not be deflected by the charge on the nucleus, since by definition, the neutron has no charge, it is electrically neutral, hence the name neutron.

Szilard knew that uranium was the last stable nucleus in the chemical table of elements. The idea Szilard had was that if a neutron was shot at a uranium nucleus, it would become unstable, and might split into two smaller nuclei. Since he knew that smaller nuclei contain fewer neutrons, that would mean that at the split, several neutrons would be shot out that could be used to split other uranium nuclei.

Hence he was the first person in the world to conceive the idea of a nuclear chain reaction. Being a physicist and a friend and colleague of Einstein, he was able to calculate how much energy would be released if such a chain reaction were to occur. He knew that the sum of the masses of the two split nuclei and the emitted neutrons would be less than the original mass of the uranium nucleus and the incoming neutron. So what happened to the missing mass?

He knew that it would be converted into the tremendous energy of the nuclei flying apart. How much energy would be released if a melon-sized mass of uranium could chain react? He realized, once he performed the calculation, that there would be enough energy to vaporize a *whole city*.

Szilard was a Jew, spoke German, and had read the book that Hitler wrote in prison after his aborted Munich putsch, "Mein Kampf" ("My Fight"), so he knew that Hitler had an "Endlosung" ("final solution") for the Jews, i.e. he wanted to wipe them out. In 1933, Hitler came to power in Germany.

Szilard was thus not only the first person in the world to conceive the notion of a nuclear chain reaction, but was also the first person to fear that Hitler might be the first to "get the bomb." This idea terrified him. Germany was the dominating country in physics in the 1920s – Einstein, Planck, Heisenberg, etc, so he felt that there was a very real possibility that Germany would be the first.

He rushed to the US and Washington DC, and started to rattle people's cages in the Pentagon, etc. The Pentagon types thought he was a loon. The concept of a single bomb being capable of destroying a whole city struck them as being ludicrous. In the

1930s, the biggest bomb could destroy perhaps a large building.

Szilard's aim to see the US be the first country in the world to develop the nuclear bomb was not succeeding, so he changed his tactics. His next idea was to meet his old friend Einstein, who by then was living in the US. Of course, Einstein understood the theory instantly, since he had invented most of it, especially his famous equation $E = mc^2$ that Szilard used to make his energy calculation. Einstein was also German and a Jew, so knew all too well what the consequences would be if Hitler got the bomb first.

Szilard had Einstein sign a letter that Szilard had drafted, that was then sent to the president of the US, namely F.D. Roosevelt, who authorized the establishment of the "Manhattan Project" that built the nuclear bomb within a few years. It was a mere 12 years between Szilard's original idea of a nuclear chain reaction and Hiroshima.

I really admire Szilard. I believe that history has undervalued him. In my view he was one of the greatest unsung heroes of the 20th century and whose historical reputation deserves to be far greater than it currently is.

Now I would like to make my point. Szilard was alone at first, before he began to persuade other physicists of the correctness of his vision, namely that future nuclear physics would soon be capable of building a single bomb so powerful that it could flatten a whole city. He knew that he was right, he knew the theory, and he persisted in persuading the powers of the time to take action.

Something similar is starting to happen now. A growing number of researchers and professors in the field of artificial intelligence are seeing the writing on the wall, i.e. that 21st century technologies will make very real the possibility of building godlike massively intelligent machines. These people have the theory.

They see what is coming, and they are starting to get worried. I am one of them. This book is an attempt to raise the alarm on this issue. What should humanity do when/if humanity has the technology to build artilects?

My point here is that if you find the whole notion of artilects too science fiction like, then consider what Szilard was saying in the early 1930s. There is quite a close analogy here. Imagine how crazy the notion of a city-destroying bomb must have appeared in the 1930s, yet it became a reality. If the notion of building machines trillions of trillions of times smarter than human beings seems ludicrous to you now, just remember Szilard and his predictions.

This introductory chapter has given you an overview of what the "Artilect War" is about. The later chapters will provide greater detail on the ideas sketched out so far.

I hope this book will make you think. It is written to help make you conscious of an issue that I believe will dominate the global politics of the 21st century, that will color and define the age, namely, the question of "species dominance." "Should humanity build artilects or not?" This question I believe will divide humanity more bitterly in the 21st century than the question which divided humanity so bitterly in the 20th, namely, "Who should own capital?" The bitterly opposed answers to that question led to the Capitalist/Communist dichotomy. The question that will dominate 21st century global politics, I believe will be, "Who or what should be dominant species on the planet, artilects or human beings?"

I end this chapter with a little slogan that expresses rather pithily, the essence of the artilect debate :

"Do we build gods, or do we build our potential exterminators?"

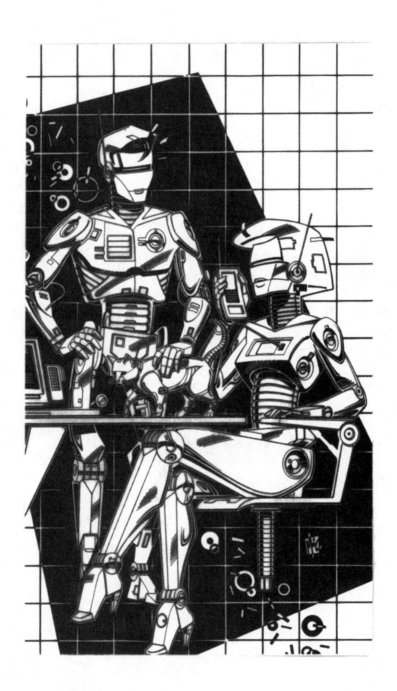

Chapter 2

The Author and His Work

Who is this *de Garis*, who makes such outrageous claims -- that machines will become trillions of trillions of times smarter than human beings by the end of the century -- that there will be a major war over the issue of species dominance, and that as a result billions of human beings will die? Is he a mad man? Is he a science fiction writer? Does he deserve to be listened to, or can humanity afford simply to ignore him?

In an attempt to establish the credibility of the author, and his ideas, this chapter will be divided into three main sections. The first gives a brief autobiography, the second is a longer description of his current work, and the third is a presentation of his future work goals and dreams.

2.1 Autobiography

Switching to first person, I was born in Sydney, Australia in 1947, making me a middle-aged man at the time of writing. I've been divorced, having had two children by my first wife. I remarried and was then widowered. By temperament I am a passionate intellectual, with over 6000 books in my private library. I am a scientist, a research professor, a social activist, a writer, and a social critic.

As an adolescent, growing up in Australia, I felt that my

27

passionate intellectual values were not valued by Australia's phlegmatic anti-intellectual brawn-based culture. During the time of the Sydney Olympic games, a BBC journalist said of Australians that they would rather win a gold medal than a Nobel Prize. By the time I was 23 and had finished my basic undergraduate degrees, in applied mathematics and theoretical physics, I wanted to leave the country forever.

The first day in London I felt overwhelmed by the feeling that I had set foot in an unquestionably more compatible and sophisticated culture. That night on BBC TV, watching a debate, I was struck by its intellectual quality. I felt a great weight lift off my shoulders. I had found my home, a culture that valued my values.

After arriving in the UK, and spending a year in London, with its awful air pollution in the early 1970s, I had constant catarrh and decided to move to beautiful and academic Cambridge. I became a freelance mathematics tutor to the undergraduates of some half dozen Cambridge University colleges. The students would come in pairs to my apartment and be helped with the problems they were having with the math questions given to them by their lecturers.

After a few years in Cambridge, I was browsing a world atlas that I had bought for my first wife, an Australian, whom I had met on the 5-week boat trip from Australia to England. The idea occurred to me that I could live in a cosmopolitan city like Brussels and hence benefit from the intellectual stimulus of several sophisticated cultures. All I would need to do would be to learn a few languages and then move there.

Despite the prospect of living the cosmopolitan life in the future, I loved the four years that I lived in Cambridge, with its green, its beauty, its academic traditions and especially its intellectuality. It was one of the happiest periods of my life. But eventually, I had to move, because there were no long-term employment opportunities there for me.

So I moved to Brussels. I became fluent in French, German and Dutch, as planned, and absorbed those cultures into my personality, making me a much richer person, a "multi" (i.e. a multi-lingual, multi-cultured person) as distinct from a "mono" (a mono-lingual, mono-cultured person). As a "multi" I found the company of other multies far more stimulating than that of monos.

I loved my new life in Brussels. Unfortunately, my Australian wife did not. She longed to get back to her native cultural and linguistic roots in Australia. She pined. This conflict of interests eventually broke us up. She took the kids and settled back into Australia.

After the breakup I lived with, and later married, a French speaking Belgian woman. Not surprisingly, my spoken French improved rapidly.

I got a job at a large Dutch electronics/computer firm but became increasingly frustrated and bored. I missed the intellectual life of Cambridge with its academic lifestyle. After working for a few years in the computer industry, I began a PhD in artificial intelligence and artificial life at the University of Brussels, and became a researcher.

Early in 1992, my second wife and I left Europe to live in Japan. I was offered a postdoc fellowship to do AI in Tsukuba's "Science City." I believed at the time that by the year 2000 Japan would be the world's dominant economic power, overtaking the US. It was not to happen. I spent eight years in Japan, working towards building the world's first artificial brain.

Japan was too feudal, too fascist, too repressive of

29

individualism, too intellectually sterile, too socially backward for me to tolerate it for very long. I stayed as long as I did because at least Japan was paying for the construction of a remarkable new type of computer that I will discuss further in the next section.

I got a new job doing the same kind of work at a research lab in Brussels. I returned alone, because my second wife had by then died of lung cancer. She had smoked heavily until I met her, stopping at my insistence, but I suppose the damage had already been done.

The Brussels lab, privately owned, bought one of my brain building machines (one of four in the world, at a price of $0.5M each). The lab was founded during the dotcom boom, so I decided to invest $100,000 of my Japanese savings into it, hoping to become a millionaire.

My brain building work and my machine attracted worldwide media attention -- I was getting two international media contacts a week. France's elite newspaper "Le Monde" wrote a dozen articles on the artilect issue and attracted enough attention to generate a political senate hearing. France was the first country in the world to address Cosmism at the political level (Paris, July 2001).

It looked as though life was going swimmingly until disaster struck. The dotcom boom turned into the dotcom crash. Investors stopped investing their money in hi-tech blue-sky research labs, and I lost my $100,000 when the lab went bankrupt, as well as my job.

My next job was as a professor of computer science in the US. Ironically, my first working day in America, happened to be on Sept. 11th, 2001. My head of department met me that morning at the university hotel, saying, "Hi Hugo – have you seen this?" pointing to a TV set. "Odd behavior," I thought, but looked politely in the direction he indicated. I saw a building on fire, and then recognized it as NY's World Trade Center, and the penny dropped. Some time later in the university's cafeteria, eating breakfast, we heard a student scream out, "They hit the other one!" "Is America always like this?" I thought.

I then had to adapt to the individualism and laissez-faire attitude of the US as a professor, adjusting to the needs of teaching students, fighting like crazy to get research grants and adapting to my 6th country. It all took a lot of nervous energy, so I did not push my Cosmist ideas in the US media.

A few years later, I gave a talk to some retirees on the artilect debate. When I finished, I was approached by a publisher who asked me, "Have you considered writing a book on your ideas?" The answer to that question lies literally in your hands dear reader.

2.2 My Work

In this section I will describe at some length the work that I have done over the years, with emphasis on what I have been doing during the past decade, since it is most relevant to the theme of this book.

Starting in the late 1980s, I began to evolve neural networks using a form of software simulated Darwinism, called Genetic Algorithms (GAs). I began publishing a steady stream of scientific research articles. I had 20 published by the time I had finished my Ph.D.

A neural network can be envisaged as a 3D array of brain cells -- neurons -- interconnected by branch like fibers called axons and dendrites. In an axon, a signal originating from a neuron travels away from it. In a dendrite, the signal is sent to the neuron. When an axon connects with a dendrite or another neuron, the junction is called a synapse.

In a real biological brain, each neuron or brain cell can have tens of thousands of synapses, i.e. it can be influenced by signals arriving from tens of thousands of other neurons. Those neural signals arriving at the neuron at the same time get reinforced or "weighted" and then summed. If the total signal strength is above the threshold firing value of the neuron, it will fire, i.e. it will send pulses of electricity down its axon at a frequency proportional to how much greater the summed value is above its threshold value. The axon pulses then travel down to their synapses to influence further neurons.

This kind of biological neural network can be simulated in software. Typically the number of neurons simulated in a single network in the 80s and 90s was tens to hundreds. For my Ph.D. work, I was using at most 16 neurons per network. This contrasts so sharply to my present work, which deals with nearly 100 million artificial neurons.

The next few pages describe the work I do in more detail, and are somewhat technical. I ask you to bear with me if you find their understanding somewhat difficult. If you don't understand them, skipping this section will not unduly disturb the general flow of the book. Consider also, that this book contains an extensive glossary that may prove helpful from time to time with unfamiliar terms.

A genetic algorithm (GA) uses a software-simulated form of Darwinian evolution to optimize the performance of whatever is being evolved. For example, take my application of GAs to the evolution of neural networks. I simulated the behavior of a neural net in the following way. The first problem was how to represent the neural net itself. I took 16 neurons and had them all connect to themselves and all other neurons, so that there was a total of 16*16 = 256 connections. The incoming signal strengths, represented by ordinary decimal pointed numbers, e.g. 10.47, were multiplied by a weighting factor, e.g. 0.537, and then summed. As an illustration of this idea, imagine a very simple network of only 2 neurons, hence 4 connections. Neuron 1 sends a signal to itself at connection or synapse C_{11} and to neuron 2, at connection or synapse C_{12}. Neuron 2 sends a signal to itself at connection or synapse C_{22} and to neuron 1 at connection or synapse C_{21}. Assume that the signal strengths at a given moment are S_1 and S_2 (e.g. 10.54 and 7.48).

Each connection C_{ij} or synapse possesses a corresponding weighting factor w_{ij} that is used to multiply the signal strength of the signal coming through the synapse. So the sum of the signal strengths arriving at neuron 2 would be $(w_{12}*S_1 + w_{22}*S_2)$. Similarly, for neuron 1. There will be 4 of these weights. Assume that the value of each weight lies between -1 and $+1$. Thus each weight can be represented as a fractional binary number with say 8 bits (binary digits, 0s or 1s). Four such numbers can be represented by 4*8 = 32 bits that can be laid out in a row of length 32 bits. With 16 neurons, I had a row or bitstring as it is called, of 16*16*8 = 2048 bits to represent the 16*16 weights of my neural network that I was evolving.

If I knew the 2048 bit values (0s or 1s) I could calculate all the 256 weight values, and hence construct a fully interconnected neural net from them. The reverse process was also possible. If one knew the values of all the weights, and the values of the initial incoming signals from outside the net, one could calculate the signal strength of each neuron as it fired. If one knew how each neuron fired, one knew how the whole neural network fired, or behaved. One could extract the signals from some of the neurons and use them as control signals to control some process, e.g. the angles of legs of a robot to make it walk.

To explain how to use a GA in this application, imagine generating 100 random bitstrings of length 2048 bits each. From each bitstring one can construct its corresponding neural net. To each net one applies the same initiating input signals to kick-start the signaling of the network. One extracts the output signals and uses them, for example, to make some stick legs walk by controlling the angles of the 4 lines that make up the stick legs. One then measures how far the legs walk in a given time.

Those bitstrings that generate neural nets that generate a longer walking distance, survive into the next generation. Those that generate a shorter distance walking are killed off, Darwinian style, i.e. "survival of the fittest." The fitter bitstrings, i.e. those with higher performance scores or "fitnesses," have copies made of them, called their "children" or offspring. The children and their parents are then "mutated," meaning that at low probability, each bit may be flipped (a 0 to a 1, or a 1 to a 0. Two bit-strings can be "sexed," a process called "crossover." There are various ways to do this. One simple way is to take two parent bit-strings or "chromosomes" as they are usually called, to cut them both at the same position, and then swap components. This is the equivalent of sex, which is basically only the mixing of genes from two parents to form the offspring.

The fitter parents have more offspring. Each generation of the GA has a fixed population size, e.g. 100. Most mutations and crossovers cause the chromosomes to have lower fitnesses, so they get weeded out of the population. Occasionally, a mutation or crossover actually improves the fitness of a chromosome by a small amount, so that in time it squeezes its parents and other inferior chromosomes out of the population. By looping through this cycle hundreds of times, it is possible to evolve a neural network, or whatever one is trying to evolve, which performs quite well.

For my Ph.D. at the University of Brussels I was evolving neural networks that gave time dependent output signals. As far as I know that was the first time that anyone had done such a thing. Previously, a few people had applied GAs to neural network evolution, but the applications were static, i.e. the signals being extracted did not change with time. This struck me as being unnecessarily restrictive. The GA should be able to handle time dependent outputs. Once I had this insight, I started to evolve a neural net that made some stick legs walk. It worked. It required a few tricks to get it to evolve, but it did work.

That initial discovery, that it was possible to evolve neural network dynamics (as distinct from statics) opened up a whole new world for me, and created a new research branch called "evolutionary neural systems." I began to wonder what I would do next. The thought occurred to me that if I could evolve one behavior with one neural net, I could evolve a different behavior with a second neural net, i.e. one with a different set of weights. The weight set determines the dynamics of the output signals.

I became more ambitious. Instead of playing with simple stick legs confined to a 2D plane, I would evolve behaviors for a 3D simulated quadruped creature that I called "Lizzy." If I could evolve one behavior successfully, then I could evolve a whole library of behaviors, with one neural net per behavior. I could probably then switch behaviors by having Lizzy at first walk and then turn. To achieve a smooth behavioral transition, all that was necessary was to switch off the inputs to the "walk straight" behavior-generating network (or module as I started calling them), and input the outputs of the walk module to the turn module. Simulation experiments showed that the motion transition was smooth. Great. I now knew I could get a quadruped creature like Lizzy to display a whole library of behaviors.

The question arose as to when one would want to switch behaviors. Perhaps such decisions might arise from stimuli from the environment. I started to see if I could evolve detector modules, e.g. signal strength detectors, frequency detectors, signal strength difference detectors, etc. Yes, it was possible. The next logical step was to attempt to evolve decision type modules, e.g. of the type -- "if the strength of the 1st input signal is greater than S_1, and the strength of the 2nd input signal is less than S_2, then switch on action A_n", i.e. a stimulus signal would be sent to the module that executes action A_n.

Putting all three kinds of modules together, i.e. behavior generating or behavioral modules, detector modules and decision modules, it seemed to me that it would be possible to start making artificial nervous systems. If there were a lot of such modules, then I thought it would be fair to call such a collection, an "artificial brain." It was at this stage that I started to become very ambitious. I began to see myself as the future pioneer of artificial brains, as Mr. Brain Builder.

But there were problems. The computer I was using in the late 80s and early 90s was hopelessly slow for the task I had in mind. By the time I was playing with a dozen evolved modules, the simulation speed of Lizzy on the computer screen was becoming noticeably slow. Every time I added another module's weights, the simulation speed slowed further. It became obvious to me that this was not the way to go. How to get around this problem?

By this stage I had finished my Ph.D. and was now postdocing in Japan, in 1992. In the summer of that year I was in the US talking with an electronic engineer acquaintance of mine at one of the universities that I am associated with, namely George Mason University in Virginia. I was asking this acquaintance how it might be possible to use electronics to speed up the evolution of my neural net modules. After about an hour's discussion, he mentioned something called an FPGA (a field programmable gate array). Not being an electronic engineer, I had never heard of such a thing. "What's an FPGA?" I asked. He told me that it was a special kind of chip that was programmable, i.e. one could send in a bit string that would instruct the chip how to wire itself up (or "configure" itself, to use the technical term.)

I suddenly got very excited. A vision flashed before my eyes. Since I had spent the past few years using GAs to evolve neural nets, my immediate inclination was to imagine the configuring bit string as a GA chromosome, so the idea that it might be possible to evolve hardware directly in the chip suddenly looked plausible. I began to grill my acquaintance. Can the configuring bit string be sent in an unlimited number of times? He thought for a moment, and replied that if the chip were based on RAM, i.e. computer memory, then like ordinary RAM in any computer, the programmable chip could be reprogrammed as often as one likes.

I felt overjoyed. It meant that it might be possible to send in random bit strings that would configure or wire up the programmable chip in a random way, generating a complex random circuit. If there was another circuit, programmed by a human being to measure the performance of the randomly

37

programmed chip, then it might be possible to perform a GA directly in hardware at hardware speeds.

I was so excited by this vision, that as soon as I got back to my Japanese research group, I gave a seminar on my idea and launched the research field of "evolvable hardware." I wrote papers on this idea, preached it to colleagues, gave talks on it at conferences, etc. The research field of Evolvable Hardware, or just EH, is now an established research field, with its own conferences every year or so in the US, Europe and Japan, plus its own academic journal. I feel I am the father of this field, and use its basic ideas in my daily work.

The following year, 1993, I moved to a research lab in Kyoto, Japan where I began work on building an artificial brain. I was convinced, after my discovery of the possibility of evolvable hardware, that I had found a tool that would make the building of an artificial brain practical.

I started writing papers announcing that I intended to build an artificial brain with a billion artificial neurons by the year 2001. In 1993, to make such an announcement invited disbelief, because at the time, most neural net researchers were dealing with tens to hundreds of neurons, as I had been in earlier years. To hear someone suddenly announce that he was going to use a billion, sounded ludicrous. I was laughed at, ridiculed.

But, I was convinced then that my vision was sound. If one could build a special kind of computer based on the principles of evolvable hardware, then its electronic evolution speeds would make brain building practical. I did the math and reasoned that a billion neuron artificial brain by 2001 would be just about doable. I had a contract with my Japanese lab for 7-8 years, so I thought I had the time to be ambitious.

My first task was to choose some kind of medium in which to grow and evolve neural nets. I chose to use cellular automata (CA). Each cell of a 2D cellular automaton can be likened to a square on a chessboard, but with two differences. One is that the chessboard has an unlimited number of squares. The other is that the squares are not confined to be only black or white but can be any of a finite set of colors. Each square can change its color into any other of the set only at the tick of a clock. The color that a particular square changes into depends on its current color, and the colors of its four immediate, touching neighbor squares. For example, if the North square is red, the East square is yellow, the South square is blue, the West square is green and the central square in question is brown, then at the next clock tick, the central square changes its color to purple.

By appropriately choosing thousands of such rules, it was possible for me to make these cellular automata cells behave like a neural network that grows and evolves. For example I could grow pathways three cells wide, in which I would send growth cells that moved down the middle of the path. When a growth signal hit the end of the growing path it would make the path extend, turn left, turn right, split etc, depending on the color of the growth signal. By mutating the sequence of these growth signals that were sent down the middle of the CA pathways, I was able to evolve the CA based neural net.

This process occurred in two phases. The first was the growth phase. After a few hundred clock ticks, the growth would saturate. No more 3-cell wide CA trails or paths could be grown. These trails were the axons and dendrites of the neural net. Once the growth phase was completed, i.e. the growth instruction cells had cleared themselves from the network, the grown neural net could then be used for the subsequent signaling phase. Input signals could be applied, which would propagate over the network. These signals behaved like the signals in the neural networks that I had evolved in earlier years. They could be extracted at output points and used to control processes whose fitness or performance quality could be measured. The fitness of the performance became the fitness of the network, which in turn was grown from a sequence of growth instructions, i.e. a random string of 6 different integers (whole numbers).

What I had done was marry neural nets with cellular automata. This had not been done before as far as I know. The reason for doing this was that I thought CAs would be a suitable medium in which to have billions of CA cells, more than enough for a billion neurons. It seemed to me to be practical. The workstations (i.e. computers a bit more powerful than PCs) of the time would have a gigabyte (a billion bytes) of RAM memory in them. RAM is cheap, so since I could store the state or color of one CA cell in a single byte (8 bits) of RAM, and my workstation could have a gigabyte of RAM, that would allow me to store the colors of a billion CA cells, a billion! That's a lot, more than enough in which to put an artificial brain with a huge number of neurons. Space would not be a problem. The technology of the time would allow it. It would be practical.

It took me about a year to write all the rules (North-East-South-West-Center type rules) to show that a 2D version of a CA based neural net would work, that it would evolve. I had to hand code (with software productivity tools to help me) about 11,000 such rules to get it to work, but work it did. I successfully evolved oscillator circuits, signal strength detection circuits, line motion detector circuits etc. It was time to move on to a 3D version that would have quite a different topology. In 2D, circuits have to collide. They cannot go past each other. Whereas in 3D, CA trails can pass each other using the 3rd dimension. The dynamics and evolvability of 3D circuits would be much richer than the 2D case. I got the 3D version to work but only after another 2 years, and roughly 60,000 rules.

By this stage I was feeling quite miserable in Japan. My immediate group boss had a policy of having only one person per project, which made me terribly lonely and intellectually sterile. I had no one to really talk with. After exerting some pressure I finally got a young German M.Sc. level student to help me for the year 1996.

I explained to him that the 3D version was pretty well finished, and that I was becoming increasingly disillusioned with the particular CA model that I had been using. I explained to him my dream of growing and evolving CA based neural circuits directly in electronics, at electronic speeds. I felt that together we would need to simplify the CA model, so that it would be possible to fit it all into the electronics of the time, i.e. 1996. He listened to my list of desiderata and then disappeared for two weeks. He returned with a new, much simplified neural net model that kept the essential features of my old model, but added features that simplified it to such an extent that indeed the new model could be put directly into electronics. This new model was called "CoDi."

At about this time, in the second half of 1996, I was contacted by an American electronic engineer. He had found my papers interesting and wanted to collaborate. I sent him details of my German assistant's new model and asked him if he thought he could implement it in hardware using special FPGAs that were then on the market. He said he thought he could. My Japanese boss

41

approved the financing of the idea and a close collaboration between my new American colleague and myself then started. Unfortunately I lost my German colleague only after one year. He went to do a Ph.D. in Europe. My Japanese boss reverted to his old policy of one person per project and I became more miserable than ever.

I was starting to want to leave, but couldn't, because the new machine had just been approved for construction. Relations with my group manager became increasingly strained, especially after I discovered he had a policy of putting his name on academic journal papers written by his subordinates to which he made absolutely no intellectual contribution. He asked me to put his name on one of my journal articles. I refused, and told him that in the west that would be considered disgusting, an abuse of power, and corrupt. After that, relations soured fast. I was allowed to stay on until the end of the year 1999, and then I would have to leave. The Japanese economy had performed so badly during the 90s, "the lost decade," that the whole research division was considered too blue sky, too fringy to be funded in times of economic scarcity. So I and most of my department left Japan at the turn of the millennium. I got another job. Ironically it was in Brussels, and to do the same work as I had done at my Japanese lab.

During the years 1996 until 2000, my hardware colleague in the US was working away solidly on constructing the special piece of hardware that would fulfill my dream of building artificial brains. It was slow going for him. He had only a limited budget from my Japanese group boss. He could afford only one full time assistant plus a few part timers on limited term contracts.

During the course of his work, the US company making the FPGA chips that the machine was based on, decided to take them off the market. My US colleague then had to fight the company to get the remaining chips. This caused many months of delay. The chips were finally obtained, but were untested. Thus he had to test them himself, without the thorough testing software that the company would have – more delays.

It was not until mid 2000 that the machines that I called CAM-Brain Machines (CBMs) were finally debugged sufficiently that true evolution experiments could begin. CAM stands for Cellular Automata Machine, because the original work was to put an artificial brain inside cellular automata.

The first CBM was delivered to my Japanese lab in early 1999, but it still contained bugs. With untested chips and small manpower, work progressed slowly. But all was not gloomy. Other people became interested in the CBM. By early 2001, there were four such machines in the world. The first remained at my old Kyoto lab in Japan. The second was bought by a Belgian speech-processing lab and later transferred to a bio-informatics company also in Belgium. The third was bought by my Brussels lab, and the fourth was owned by my hardware colleague. Thus with two of the four machines in Belgium, Belgium became in a sense the world leader in this field. In 2000, I managed to get a $1,000,000 grant from the Brussels government to build an artificial brain to control a small kitten-like robot, giving it hundreds of behaviors. As you can probably see, all this work from the 1990s was really only an extension of my old Ph.D. thesis work of the 1980s.

Just what could the CBM do? I believe it was truly a miraculous machine that in time, once people appreciated its significance, would take its place in the history of computing. It implemented the CoDi CA based neural net model directly in electronics. It evolved a neural net in a few seconds, i.e. it

43

performed a complete run of a genetic algorithm, i.e. tens of thousands of neural circuit module growths and fitness measurements. It could change the color of its CA cells at the phenomenal rate of about 130 billion a second. It could handle nearly 100 million artificial neurons. It had the processing capacity of about 10,000 PCs, so was definitely a supercomputer but cost only $500,000.

The CBM had two main roles. The first was to evolve individual neural circuit modules, or just modules, I called them. A neural net was grown/evolved inside a 3D CA space of 24*24*24 CA cells or little cubes. About 1000 neurons could fit inside this space. Branch-like axons and dendrites grew randomly inside this space. A programmed FPGA was used to measure the quality of the neural signaling of the network that was grown. The basic ideas were similar to what I was working on before 1996. Once a module was evolved, it was downloaded into a gigabyte of RAM memory. 64000 of such modules could be evolved, one at a time, each with its own fitness definition (i.e. task or function) as specified by human "evolutionary engineers" (EEs) and downloaded into the RAM. Later, "brain architects" (BAs) interconnected the downloaded modules using software to form their humanly specified artificial brain architectures to perform the tasks that they wanted.

After the bankruptcy of my Brussels lab in the dotcom crash, and the move to my US university (Utah State University), I was forced to rethink. My hardware colleague lost $300,000 of his own money when my Brussels lab was unable to pay for the CBM it bought from him, making him wash his hands of the whole project. Since he had monopoly knowledge of the detailed architecture of the CBM, further work on its development came to a screaming

44

halt. My American university could not afford $0.5M for a fifth CBM, so I was left without my machine. I spent 2+ years learning how to teach, plus writing papers on evolvable hardware and quantum computing, until I realized that thanks to Moore's Law, I could once again build brains, but this time far more cheaply.

A British company had developed a way to translate ordinary computer software code (e.g. in the computer language "C") into bit string instructions to wire up (i.e. configure) programmable chips (FPGAs). I conceived a new brain building research project. The new approach was to use a GA to evolve neural nets that would be programmed into the FPGA electronic board (costing less than $1000). This board could evolve a neural net circuit module dozens of times faster than a PC could in software (the latter taking hours to several days per module).

These modules would be evolved one by one in the hardware and the result downloaded into a PC. Each module would have its own evolved function as specified by human BAs (brain architects). Once several 10,000s of these modules had been so downloaded into the PC, special software could be used to specify the connections between the modules, e.g. the output of module M3728 could be connected to the second input of module M9356.

The PC is then used to perform the neural signaling of the whole artificial brain in real time (i.e. 25 neural signals per neuron per second). Today's PCs can signal 10,000s of modules in real time. The whole approach, that I call "Brain Building on the Cheap" costs under $2000, so I'm hoping the idea will spread to other universities and research labs. Of course my approach will be a lot more persuasive to my colleagues and funders if I can actually build such a brain in the next few years and show it controlling a robot to perform useful tasks.

45

2.3   Future Tasks and Dreams

If those people who had laughed at my preposterous assertion that I would build an artificial brain with a billion neurons by 2001 were able (hypothetically) to see the CBM in 1993 as it existed in 2001, they would not have laughed. Admittedly the machine did not handle a billion neurons. The actual figure was 75 million, but that's only one order of magnitude off. That's not bad. Admittedly also, the task of architecting the artificial brain, a huge task, still lay ahead. There was still a lot of work to be done, and I still suffered from critics. With all the delays, whether for commercial, intellectual, managerial, or personal reasons, I did not have an artificial brain to show off to people. Some journalists started to get impatient, and were wondering if and when I would deliver. Since the Brussels lab was going bankrupt, I was not free to tell the journalists the reasons for the delays. It was very frustrating.

The plan was, just before the bankruptcy, to complete the machine's evolvability studies, evolving one module at a time. If the evolvability levels were not sufficient, we would have to change the fitness definitions we used in the CBM. We may also have had to change the neural net model implemented in the reprogrammable FPGAs. Once that stage could be completed, the next would have been to start building multi-module systems, with 10s of modules, then 100s, then 1000s, up to 64,000, to build an artificial brain aimed at controlling the behavior of robots. We intended to show off a kitten robot with many behaviors controlled by an artificial brain. One would not have needed to have a Ph.D. to understand what was going on, as was the case with the CBM, but just by simple observation of the robot, one would have been able   to   see   that   "there   was   a   brain   behind   it."

In parallel with all this work, which should have taken several more years, was the need to start serious thinking about the next generation of brain building machine, that I called the BM2 (brain building machine, 2nd generation). I started collaborating with another American colleague who had some revolutionary ideas for the next generation of electronics that self configure. He estimated that with a budget of a few million dollars, it would be possible to build a next generation machine within about four years, which should be about 1000 times more performant compared with the CBM.

However, since I have not yet managed to persuade US grant givers to part with such sums, I have had to be content with the more modest approach described in the previous section (i.e. "brain building on the cheap.")

In fact, it is my general ambition to continue trying to build a new generation brain building machine and its corresponding brain every 4-5 years. I'm now in my late 50s, so if I choose to retire in my 70s, that gives me about 15-20 years, or three more generations. In 20 years, if Moore's law continues to be valid that long, it will give humanity the ability to put one bit of information on one atom. Once that happens it will be possible to build what I call "Avogadro Machines," i.e. machines with a trillion trillion components. Avogadro's number is the number of molecules in an object of human scale, e.g. an apple in one's hand.

If the second-generation brain-building machine can be funded and can be built within the next 4-5 years, it will be possible to make the next generation brain more similar to the biological brain. The neural net models it implements can be more sophisticated and closer in their behaviors to those of biological neurons.

Within a mere 20 years, i.e. my own working lifetime, I and other brain builders will have the technologies and the tools to build ever more performant artificial brains.

Is it any wonder then, that I and others are beginning to feel alarmed at the rapid progress that brain building can be expected to make in the coming 20 years. What will our artificial brains be doing for humanity by that time? I would say it is highly likely they will be in our homes, cleaning them, babysitting our kids, talking with us, giving us infinite information from knowledge banks all over the planet. We will be having sex with them, be educated by them, be entertained by them, made to laugh by them etc. The brain building industry 20 years from now I estimate will be worth about a trillion dollars a year worldwide. Within a few years I hope and expect that if my own group can "prove concept" that brain building is doable, then a new "brain building" research field will be established.

If we have all this within 20 years, where will humanity be in 50 years, in a 100? Given the exponential progress in the accumulation of our knowledge of brain science, all of which can be immediately incorporated into neuro-engineering the moment it is discovered, I feel that the initial positive feelings about artificial brains will later turn sour and develop into fear.

I would like to be considered the father of the artificial brain. I feel I am already the father of evolvable hardware and of evolutionary engineering, which are the enabling technologies of this new field. If I were a traditionally minded engineer or scientist, I would probably be quite content to get on with my work and not worry about its longer term social consequences, but I'm not like that. I'm a very political animal, and I'm very worried. My rather unusual combination of being a scientist/engineer and at the

48

same time a social critic and media person makes me an appropriate person I believe to raise the alarm on the artilect issue.

I'm hoping that my credibility or otherwise as a professional brain builder will aid my attempts to raise the alarm on the rise of the 21st century artilect. However, the two need not be connected. Even if I fail to build an artificial brain, others will succeed. For me to succeed with each brain-building-machine-generation, and the building of its corresponding brain, I will need to raise more money, hire more people as the scale of the enterprise keeps increasing. In theory, I would need to become like Goddard, the US rocket pioneer, or Werner von Braun, who put an American on the moon. Both these men started with toy rockets, but had a vision. In the 1920s, Goddard's first contraptions were not much better than 2m tall ancient Chinese style rockets. Twenty years later, both he and von Braun were heavily subsidized by their respective governments to build highly sophisticated rockets capable of traveling great distances. In the late 1960s von Braun played a major part in NASA that put Armstrong on the moon.

I have similar dreams. I dream of national projects paying billions of dollars to build artificial brains. I have talked of the J-Brain Project (Japan's national brain building project), the A-Brain Project (America's), the E-Brain and C-Brain Projects (Europe's and China's). Within 20 years, and in possession of Avogadro machines, there will be a huge amount of work to be done in building a brain, not with just billions of components, but trillions of trillions of components. A huge team of people will be needed. That's my longer-term dream, 20 years from now.

After that, once I have retired, I hope I will be able to play the role of the wise old man who advises younger minds on where the whole brain builder effort ought to be headed. As this book shows,

I am not optimistic about the future survival of humanity when faced with machines that become ever smarter at exponential rates.

My ultimate goal is to see humanity, or at least of portion of humanity, go Cosmist and to do it successfully by building truly godlike artilects that tower above our puny human intellectual, and other, abilities. That is my true goal. I won't live to see it unfortunately. True artilects won't be built within the 30-40 years I have left. I will not live to see the ultimate fruits of my work. This is a source of great frustration and disappointment to me, but there is one consolation. At least I will probably die peacefully in my bed of old age. As this book shows, I fear for my grand children whom I believe are likely to be destroyed in a gigadeath war over the issue of species dominance late this century.

So, dear reader, you have now heard the more technical side of my story, and a brief description of my life's work. Does knowing this make you feel that my political opinions concerning a possible Artilect War are more credible? Should I tell you that I am not just a Ph.D. but a professor in the US and a guest professor in China. I was also a Davos Science Fellow, the only one in Japan at the time, so I got to go to the Davos World Economic Forum to entertain the billionaires. I'm in the Guinness book of world records (p126, 2001) for the CBM. I was a guest editor of a special issue of an academic journal on "Evolutionary Neural Systems," which is usually an honor reserved for the person who is considered to be the best in the world in a given specialty by the editor in chief of the journal concerned. If a lot of people consider what I am trying to do to be crankish, then I hope it is clear that at least I am a competent crank. The point of this chapter is to try to convince you that the author of this book, the coiner of the terms artilect, Cosmist, Terran, gigadeath, etc is worthy of being listened to. Whether I have succeeded or not is for you to judge.

Chapter 3

Artilect Enabling Technologies

Some years ago, when I was trying to get an earlier draft of this book published in the US, I received an email from an American literary agent, saying that my manuscript, which she had read, was "quite well written, but 'fantastical', making it a very hard sell for publishers." Since then I have found in practice, that the greatest obstacle I have to face in trying to persuade people to accept these ideas, is their seemingly "science fiction" like character. Many of my readers seem to have great difficulty in getting their minds around the enormity of what is being proposed. For example, most people, when confronted with such concepts as "massively intelligent machines" with artificial intelligence levels trillions of trillions of times above the human level, or of an "Artilect War" killing billions of people ("gigadeath"), or of asteroid sized computers, etc, not surprisingly, their immediate reaction is one of incredulity. They will often laugh at the seeming preposterousness of these ideas. Even many of my colleagues (especially the non physicists) do not take a lot of these ideas seriously. For example, I tried several years ago to persuade the most eminent "applied ethics" professor on the planet, Professor Peter Singer of Princeton University, USA, to take up these ideas. I was trying to persuade him to write a book on the topic of "Artilect Ethics," which would deal with the huge moral and ethical issues concerned with the possible construction of artilects this century.

His reply was illuminating. I quote him from one of his emails to me. "To be blunt, I am not sure how to place you between the 'total flake' and 'genius ahead of his time' views of your ideas." This is from someone with a very open mind.

So you see my most pressing problem for the moment is one of credibility. How to persuade people that these ideas are not just a piece of non-serious "science fiction," but are very probable "future science" ideas. Admittedly, the persuasion task has become easier recently as the world media increasingly takes up the message. Before coming to the US, I was constantly in the world media, (TV, newspapers, magazines, radio, web, etc) in such countries as France, Holland, UK, Australia, Poland, etc. But even the US has been contacting me increasingly lately, without any real effort on my part.

Despite the growing credibility, there is still a long way to go, so it is essential in this book for me to try to persuade skeptics that these ideas are worthy of serious consideration, and that they should not be dismissed out of hand.

This chapter is devoted to trying to persuade you that it will be possible to build artilects this century.

The fabulous technologies that will be developed in the next 100 years will be so capable and so fantastic, that they will force the issue as to whether artilects should be built this century or not.

Once you have read this chapter, I hope you will be left with the strong impression that the artilect's potential intelligence is truly gargantuan. Artilects will have the ability to surpass human intelligence levels by many orders of magnitude, not just ten times smarter, or a thousand times or even a million times smarter, but by trillions, quadrillions, quintillions, truly zillions (using the generic term) of times smarter. (If something is 10 times larger

52

than something else, it is said to be an order of magnitude larger. If it is 100 times larger, it is two orders of magnitude larger, etc).

This chapter will try to persuade you that these numbers are not exaggerated. There are very good reasons, based on the new technologies, that we expect to be developed this century, to motivate us to believe that artilect building is a realistic proposition within the next 100 years.

This chapter will be amongst the most complicated of the book, since it will be discussing scientific ideas and technologies that are new or do not yet exist. I will try to make this chapter as easy to understand for the general, non-scientific reader as I can.

As I wrote in the introductory chapter, one of my life goals, besides building artificial brains, is to raise the alarm on the "Artilect Issue," or if you prefer to call it, the "Species Dominance Issue," or the "Cosmist-Terran Conflict." There are several ways to label the same basic problem that is coming.

This issue is far too important to be confined to intellectual discussions amongst a bunch of "nerdy scientists." In time, it will concern everyone, because if Cosmists are serious in their "threat" to build artilects, everyone will be affected, one way or another. One does not initiate a great public debate by confining ones worries to the scientific specialists, a tiny proportion of humanity, less than 1%.

An effective beginning to getting people to talk about the artilect issue is to write a book. A book will help the journalists become familiar with the problem, and they in turn will write about it for the greater reading public. Similar reasoning applies to the TV and radio journalists, who can present these ideas to an even wider audience, because unfortunately, only about a half of the population reads books.

Probably the most effective way to get the message across would be to have Hollywood make a blockbuster movie on the theme. I hope this will come. A start in this direction has been made. Several filmmakers in various countries have already made documentaries about me and my ideas.

Before launching into details of these new or yet to be developed technologies, I ought to say a little about the category of readers who could most benefit from this chapter, which I think is amongst the most difficult of the book. I believe if you have studied at least high school science, you should be able to follow most of what will be described in this chapter.

To fully appreciate this chapter, I need to talk about some very "high-tech" technologies and even technologies that don't yet exist, so I will have to go into some level of detail. I hope that few readers will be put off.

If you are, I suggest you just read as much of this chapter as you can follow, without too much effort, then skip to the next chapter, which discusses the many points of view of the Cosmists. However, if you do decide to skip this chapter, I suggest at least you accept its main conclusion, which is (to labor a point) that this century's technologies will enable the building of artilects, which could become zillions of times smarter than human beings.

Moore's Law

I begin the introduction of the artilect enabling technologies of this chapter with the phenomenon known in the electronics world as "Moore's Law" that I discussed briefly in the introductory chapter. This time however, the concept will be treated in greater detail. Gordon Moore is a person, still alive in the early 21st century, who was one of the co-founders of the "Intel" microprocessor company, in Silicon Valley, California, USA. In the mid 1960s, he noticed that integrated circuits were increasing their speed and density (i.e. the number of transistors crammed onto the surface of a silicon chip) by a factor of two every year or so. This doubling rate has remained more or less true for the past 40 years and many people believe that it will continue right down to the molecular scale.

What is the point of trying to make electronic components smaller and more densely packed? If two electronic components have to signal each other, and given the finite speed of light (i.e. the maximum speed with which electronic components can send messages to each other) then the closer these components are to each other, the faster they can influence each other. Also, the smaller the size of the components, the greater is the number of them that can be crammed into a given surface area. Hence the chip can deliver greater performance because it has more components to do more things.

The microchip industry is thus under constant pressure to scale down, to make its transistors smaller, its circuits smaller. If a company falls behind in this frenetic race, it will lose sales and go out of business. If the rival company down the road is six months ahead in its development cycle, and thus releases a new batch of products ahead of your company, you are at a great disadvantage. New generations of chips and the computers based on them come out every year or two. We are getting used to this now. We know that if we wait six months or a year, we will be able to buy a better, more performant computer for about the same price.

Moore's Law is probably one of the most important technological and economic phenomena of our times. It is fueling the digital revolution, which is now driving our global economy. So many jobs and such a large proportion of the GNPs (Gross National Products) of many nations are now tied up with the electronics, computer, telecommunications industries that if Moore's Law were ever to stop, humanity would be in for a real shock. However, there is a problem.

As the size of electronic components, particularly transistors, gets smaller and smaller, a scale is eventually reached which is so small that a different set of physical laws, which governs their behaviors, begins to apply.

If Moore's Law can continue right down to the molecular scale, i.e. if the size of electronic components can reach that of molecules and still be functional, then new laws of physics must be applied. The old "classical mechanics" discovered by Newton in the 17th century is no longer appropriate, and must be replaced by the newer 20th century based "quantum mechanics."

Quantum mechanical laws govern the behavior of atoms and molecules (and even smaller scales). For example, as the line widths of wires connecting electronic components on the silicon surface of a chip are reduced below about 0.1 micron (a micron is a millionth of a meter, about the size of a bacterium), quantum mechanical phenomena begin to appear. These phenomena make themselves felt with such a strength that the usual transmission of electrons (i.e. electric current) down the wire at larger scales, is severely disturbed.

There are many other similar reasons why researchers in electronics are worried today. They know that they must shift away from conventional electronic principles into quantum mechanical principles if electronics is to continue its incredible "Moore doublings" phenomenon. Instead of looking upon these quantum effects as a disturbance of conventional electronics, a growing number of electronics researchers are accepting the inevitable, and have started to think of new electronic and computing techniques which embrace the quantum phenomena as their functioning principles.

If Moore's Law continues unstopped until 2020 or thereabouts, it will be possible to store one bit of information (a zero or a one, a "0" or a "1") on a single atom. An excited atom (in which an electron circling the nucleus of the atom has a higher energy than usual) could be interpreted as storing a "1," and an unexcited atom as storing a "0." The two different states "0" or "1" would correspond to the two different energy levels of the atoms.

The enormous significance of this scaling down to the atomic level is the huge number of potential electronic components one could then have in a given volume. It was the Italian chemist Avogadro in the 19th century who first estimated the number of molecules in an object of human scale, such as an apple. The number is so large that it is almost impossible for the human mind to conceive.

Avogadro's Number is 6.023 times 10^{23}, i.e. nearly a trillion trillion (a 1 followed by 24 zeros). That number is a hundred trillion times larger than the number of human beings alive on the Earth at the beginning of the 21st century.

Molecular scale electronics holds the promise of truly huge computational capacities, and all this perhaps by the 2020s. When I talk about an artilect having a potential artificial intelligence of trillions of trillions of times the human level, part of that assumption is based upon the enormous computational capacities that we will have in a mere few decades, that a future artilect could possess.

Reversible Computing

The above idea of having trillions of trillions of electronic components inside a small volume (say that of an apple, or a few cubic centimeters) contains an implicit assumption, and that is that the electronic circuits contained in such a volume would be distributed throughout that space. They would be three-dimensional circuits (3D). But today's electronic circuits are all 2D, imprinted on the surface of silicon chips. Why is this? Why doesn't modern electronics take advantage of the far greater storage capacities of 3D circuits?

The answer has to do with the problem of heat generation. The following paragraphs will explain.

For the past few decades, theoretical physicists have been asking themselves some fundamental questions about the ultimate limits of the physics of computing. This branch of physics goes under the label of "phys-comp" (physics of computation). One of the questions that has been asked in the phys-comp field is "What is the minimum amount of energy or heat that must be dissipated to perform an elementary computational step?"

If you put your hand over your PC, or if you have your laptop on your lap as I do now as I type this, you will be fully conscious that your computer is generating heat. Computing inevitably generates heat it seems, or does it?

In the 1960s, a researcher named Landauer discovered that what was generating the heat in computers was the process of "resetting" memory registers (a register is a linear storage chain of 0s and 1s), i.e. wiping out their contents and resetting them to 0s. He discovered that the heat was generated when information was "wiped out" or destroyed.

To be a bit more technical, to wipe out the contents of a register implies increasing its order, making it less random. In physics, the concept of "entropy" is used to measure how disordered a physical system is. For example, ice has a lower entropy than water, because it is more ordered, less chaotic.

One of the basic laws of physics, known as the "Second Law of Thermodynamics" is that entropy never decreases in a closed system (one where energy can't get in or out). So if the contents of a register are wiped out, its entropy, its measure of chaos, decreases, so where does the rest of the entropy go, given that the total cannot decrease? The answer is in the form of heat to the surrounding environment of the computing component.

Today's computers generate heat because we are using thermodynamically irreversible processes (i.e. we can't reverse the effects at a later time). We generate heat every time we destroy information, i.e. wipe out bits. Landauer thought that this was inevitable, because when he looked at how the computers of his time all functioned, he saw that they were full of "AND gates," and the like.

An AND gate is an elementary piece of electronic circuitry which has two input signal lines (A and B), and one output line. If both input lines are set at a high voltage (i.e. have a "1" on their line) then the output line will become a "1," i.e. if both input line A AND input line B are set at "1," then the output line becomes a "1." In any other case (i.e. A=0, B=0; A=0, B=1; A=1, B=0) the output line becomes a "0."

Since there are two input lines containing a total of 2 bits of information in an AND gate, and only one output line containing 1 bit of information, of necessity, the AND gate destroys information. (If you are told in which state a system is in, that can have two possible states, then you are given 1 bit of information. For example, take the question "On which side of the road do the Japanese drive?" When you are told "On the left-hand side," you have been given 1 bit of information).

Every time two bits go through the AND gate, only one bit comes out. The AND gate is irreversible, i.e. you cannot always deduce from the output what the input was. For example, if the output was a 1, then you know the inputs were both 1, but if the output was a 0, you don't know if the inputs were, (0,0), or (0,1) or (1,0). For a gate to be reversible (i.e. you can deduce what the inputs were from its outputs and vice versa), common sense says that you have to have the same number of input lines as output lines.

People began to dream up reversible elementary circuits (or "gates") with an equal number of input and output lines. (A "gate" is an elementary electronic circuit that performs some basic operation, e.g. an AND gate, an OR gate, a NOT gate, etc). One famous such gate was called the "Fredkin Gate," which had 3 inputs and 3 outputs. The Fredkin gate is reversible, so no bits of information are destroyed. It is also "computationally universal," i.e. by feeding the outputs of Fredkin gates to the inputs of other Fredkin gates, larger circuits of these gates can be built up that can perform any of the functions that computers need to perform.

Since the individual gates of the computer were reversible, the computer itself could be made reversible. In other words, one could input the initial bits into the left-hand side of the computer, and these would be processed by the Fredkin gates in the computer design. The resulting output (the answer) would appear exiting from the gates at the right-hand side of the computer.

You can make a copy of the answer (which might generate a

little bit of heat) and then send the answer back into the computer from right to left. Since all the gates of the computer are reversible, you will end up with what you started with at the left-hand side. You have performed a reversible computation. No bits have been lost and hence no heat has been generated. Nevertheless, you have the answer you want, because you made a copy of it half way through the computational process, i.e. before you "reversed" the direction of processing.

Reversible computing may take twice as long as traditional computing, because you have to send the result backwards through the same circuit (or an identical copy), but at least there's no heat generated.

What is the significance of this! Why am I spending so much time and energy explaining such things? Because I believe the theoretical discovery of reversible, heatless computing in the 1970s was one of the greatest scientific discoveries of the twentieth century and is of great relevance to the main ideas of this book.

Since this is such a strong statement and will probably be treated with some skepticism by many people, particularly some of my research colleagues, let me try to justify why I have this opinion.

A few years ago, some phys-comp theorists were wondering, "If Moore's Law extends right down to the molecular scale, how hot would molecular scale circuits become if one continues to employ conventional irreversible, bit destroying, information processing techniques?" The answer was shocking.

Not only would such highly dense circuits melt with the heat, they would become so hot they would explode. It became clear that molecular scale circuits, if ever they are to be built, would have to abandon the traditional irreversible style of computing, and start using the new reversible style.

Only recently have researchers started thinking seriously about reversible computer designs. The laptop and palmtop computer industries are very interested in reversible computing, because it might help them with their "battery lifetime" problem.

If their computers could use electronic circuits that were more reversible, the circuits would consume less battery energy, because they would generate less wasteful heat. Hence the battery would drain more slowly and have a longer life. Consumers will be more likely to buy laptop computers that have batteries that last longer. Wouldn't it be nice to have a single laptop battery that lasted for a full transatlantic flight, for example.

So, it is inevitable that reversible computing has to happen. As Moore's Law continues to bite, pressure will increase on computer designers to use the reversible paradigm. It is only a question of time.

But, if we start taking the concept of heatless computing seriously, we can begin to play with some revolutionary ideas. For example, why are today's electronic circuits two-dimensional? Why do we talk of 2D "chips" (i.e. slices) of silicon, rather than 3D "blocks?" Well, because of heat. If we made 3D blocks of silicon with today's level of density of electronic components, there would be so much heat, the blocks would melt. Also, how would we build them and debug them once they were built? We do not have the techniques yet to do such things. We don't even bother trying to build 3D circuits because we know it would be a waste of time, due to the heat generation problem.

But, with reversible heatless circuits, we have the luxury to build large 3D circuitry with, in principle, no limi to size. We could make circuits the size of a cubic centimeter, or a cubic meter, or the size of a room, or a house, or a building, or a city, or even a large asteroid hundreds of kilometers across. (An asteroid is a huge boulder of metal or rock that orbits the sun at a radius between those of Mars and Jupiter. There are thousands of very large asteroids in the "asteroid belt.")

In theory we could make computers the size of moons or planets, but the gravitational effects might prove to be problematic.

You are now probably beginning to suspect why I think reversible computing is so terribly important. Ask yourself how many bits of information you could store in an asteroid, for example. The answer is about 10^{40}, i.e. a "1" followed by 40 zeros, i.e. ten thousand trillion trillion trillion atoms and hence bits.

Also ask yourself how many brain cells (neurons) we have in our heads. The answer is of the order of 10^{11}, i.e. a hundred billion. If we could accurately simulate in a computer the behavior of one biological neuron using a trillion bits (and that may be overkill) we would still have 10^{17} (17 = 40 − 11 − 12, i.e. a hundred thousand trillion) *human brain equivalents in one asteroid.*

I suggest you really study these numbers. They are the writing on the wall for me. What this means is that, sooner or later, humanity will be able to create vast computing capacities, eclipsing enormously human brain levels. *It is therefore only a question of time before humanity has to choose whether to exploit fully such enormous computing potential or not.*

Let me spell out a bit more explicitly just how phenomenal such an asteroid sized computer might be, and what it could do. Firstly, it could "think" a million times faster than do our brains. The neurons in our skulls communicate with each other at a maximum speed of hundreds of meters a second. E! ctronic signaling speeds, as in computers or an artilect, would be a million times faster, i.e. close to the speed of light, which is 300 million meters a second.

Even if these artilects had the same intelligence levels as humans, they could do in seconds what takes us years. Instead of getting a Ph.D. in 4 years, it would take such an artilect only $4*50*5*8*60*60/1,000,000 = 30$ seconds. But an artilect has far more than just one human brain equivalent. It has zillions of times more. So if it could distribute its thinking over all its brain, then it could do what we do in 4 years in picoseconds or less. (A picosecond is a trillionth of a second).

Artilects would be so fast in their thinking that our human pace of thought would seem as slow to them as humans trying to communicate with rocks. Over millions of years, rocks change their shape, which might be interpreted as conveying a message, but humans don't have the patience (nor the life span) to wait.

There is a strong case to be made by the Cosmists that advanced artilects would be totally bored by humans with our glacial thinking speeds, and simply ignore us. They would invent whole histories within themselves in the time it would take us to utter one word.

But the artilects need not be limited to human intelligence levels. It is not difficult to make out an argument saying that one ought to be able to extrapolate the trend in human IQ levels as we discover the neurobiological structures and functions that make one human being smarter than another. In time we should be able to look at an ordinary person's brain and Einstein's brain and notice neuro-physiological features that correlate with higher intelligence.

We could then plot a graph depicting IQ on the vertical axis, and the neuro-physiological features that correlate with high IQ

64

(e.g. number of connections per neuron in certain regions of the brain, etc) on the horizontal axis, and then just extend the trend. We may see the development of an "Intelligence Theory," as we learn more about how the human brain works and understand just what it is that makes humans more intelligent than other animals. It may become clear to us, that if we simply increase certain parameters in the design of artificial brains, we may be able to increase the level of intelligent behavior in the robots that these artificial brains control.

So, asteroid sized artilects need not be limited to architectures that generate human level intelligence. Artilects could not only think faster, with hugely more components, but in qualitatively superior ways as well.

Their huge surface areas, would allow them to attach huge numbers of external sensors to themselves, including use of the full range of electromagnetic wavelengths from gamma rays to radio waves. They could communicate with other asteroid artilects across the asteroid belt and deeper into space.

Such asteroid sized artilects using nanotech-based principles are probably the logical extreme of human technological imagination (unless we can create something called "femtotech," which I will discuss a bit later.)

Before asteroid sized artilects are built, earlier versions will certainly be much smaller, more on a human scale, but even at this smaller scale, when talking about one bit per atom, we will still have many technological problems to solve in order to build such computers.

Nanotechnology: Molecular Scale Engineering

This brings me to the need for "nanotech," as distinct from "femtotech," which I mentioned just above. "Nanotech" is an abbreviated form of "nanometer scale technology," i.e. molecular scale engineering. Nanotech builds things at the scale of a nanometer, which is one billionth of a meter, about the size of molecules. "Femtotech" is an abbreviated form of "femtometer scale technology." A femtometer is a quadrillionth of a meter, i.e. a thousandth of a trillionth of a meter, which is about the scale of a proton or neutron inside the nucleus of an atom. Femtotech would be nuclear or even quark scale engineering. Quarks are "elementary particles" which combine to build protons, neutrons and other such particles.

The first people to think about the possibility of building things at the nanometer level were presenting their ideas in the 1950s. In the 1990s, these ideas had become well accepted and regular monthly progress was being made in this domain. The essential idea is that atoms can be placed into exact position to build molecular scale machines, e.g. tiny molecular scale robots that pick up atoms and position them to build molecular scale structures.

When one begins to imagine the kinds of things that could be done with molecular scale machines, the field begins to sound truly science fiction like, yet the possibilities exist. Many scientists believe that given the current rate of research progress in the field of nanotech, it will probably be well established by about the year 2020. This is about the same time that it will be possible to store one bit of information on a single atom, according to Moore's law.

Consider some of the more fantastic things we could do with

66

a fully-fledged nanotech. Imagine tiny robots sent into the blood stream of human beings, which are programmed to detect cancer cells. They would travel throughout the body, detect the cancerous cells, kill them, and then self-destruct or be flushed out over time. A similar story could hold for "immortality generating" robots, which could repair aged cells and restore them to a state like those of young children. With a regular dose of such "fountain of youth" robots, people could become immortal.

Each of the cells in our bodies contains a DNA program that explicitly or implicitly causes the cell containing it to die. This DNA program takes the form of a molecular structure that can be reprogrammed by a molecular scale robot, a nano device. Hence nanotech offers humanity the prospect of immortality. If that happens, we will need a new politics to decide who lives forever, who dies, and who reproduces.

Another favorite nanotech idea is to have one's head or one's whole body frozen soon after death on the assumption that in a century or so, it will be technologically possible to restore the damage to the dead brain and make it come alive again. Nanomachines, the theory goes, would be able to enter the dead tissues and repair them.

There are already hundreds of people who have paid for their bodies or heads to be frozen for an indefinite period. They establish a monetary fund whose interest pays the cost of the apparatus and materials to keep the body frozen.

Molecular scale robots (nano scale robots, or "nanots") could build copies of themselves. They could reproduce and hence grow exponentially in numbers, 1, 2, 4, 8, 16, 32, 64, etc. After 20 such doublings, the numbers are into the millions. If ways could be found to make these nanots cooperate to build human scale products, then conventional economics would be revolutionized. It would cost almost nothing to make huge numbers of nanots, which then build the product. The price of the product would then be merely the cost of the raw materials. Goods could become amazingly cheap, effectively costing nothing.

The cost of designing the first such self-reproducing nanot, capable of cooperating with others of its kind to build specific products, would be amortized over the many purchases of the same nanot design all over the planet, and hence would cost almost nothing. The whole concept of economic scarcity would need to be reconsidered. Economics as a specialty would be revolutionized.

Construction materials could be made many times stronger, because today's materials still contain cracks, faults, etc that weaken their strengths. Nanotech could assemble these materials with atomic precision with no faults, no cracks, no blemishes, and hence they would be much stronger. This would probably mean that we could construct buildings that would be kilometers high if we wanted to build them. The structural skeletons of the buildings would be strong enough to withstand the stresses and strains generated by strong winds. Diamond like materials could be built with amazing strength.

As I see it, there are at least two major paradigms conceivable when discussing how nanotech could build human scale products. One is to imagine zillions of self-reproducing nanots that, once reproduced, would combine to build the product.

How would such a mammoth task be coordinated? One would need to think of the manufacturing process like a molecular scale city with a huge infrastructure to make it all happen. One could imagine nanots each doing their tiny thing on conveyor belts, assembling their few atoms at this point, at that point, and passing

68

on the result, farther down the line, where other nanots do something different. It would be Henry Ford at the nanoscale.

With zillions of nanots doing the same thing simultaneously (in parallel, as computer people say) it would be conceivable to imagine a human scale device being built. Molecular modules could be built from atoms, and these modules used as components to build larger macro-modules, which in turn become components of macro-macro-modules etc, until a human scale product is built. To make such a construction system work, the enormous molecular infrastructure needed may or may not prove to be very practical.

Artificial Embryology

Personally, I prefer a nanotechnology based on the method nature has used for billions of years to build its life forms, i.e. an "embryological approach." In the embryological process, one starts with a fertilized cell that divides and divides until some cells (depending upon their position in the embryo) begin to differentiate. Their intercellular environment sends them chemical signals that are used to switch on and switch off certain portions of their DNA, which in turn, results in different proteins being built, which perform different tasks. These different proteins then change the nature of the differentiating cells. Eventually, the mass of differentiating cells creates a living three-dimensional biological creature.

Evolution has created a growth mechanism that takes a linear one-dimensional coded string of chemical instructions (usually called DNA) and translates it into a three-dimensional functioning living creature. The study of how this miracle of nanoscale engineering occurs is called "embryology," or "development." The

69

machines that instruct the differentiating cells how to switch on and switch off genes at the appropriate time in the growth process are of molecular scale. A biological cell can be viewed as a molecular scale city, with millions of molecular inhabitants all organized into one functioning whole.

I would like to see the creation of a new branch of science that I call "Artificial Embryology," which would aim to mimic the same process that nature employs to grow dinosaurs or gnats from single fertilized eggs. Scientists and engineers would need to understand how nature does it in far greater detail than is known at the beginning of the century. But, as the molecular biologists are discovering all there is to know, more or less, about certain single-celled bacteria, many of these scientists are changing specialties towards studying how multi-cellular creatures are built. Embryology is now a hot research topic, so we can expect a steady flow of discoveries in this domain over the coming decades.

Eventually, I expect to see the creation of what I call "Embryofacture" (embryological manufacture), i.e. using artificial embryological techniques to manufacture human scale products from the nano scale. Instead of needing a complex nanoscale infrastructure using nanots as described earlier, one would need a complex timing control system which decides when particular genes in the DNA (or its humanly designed equivalent) switch on and off when stimulated by certain molecular signals in their inter and intra cellular environment.

Designing such a complex control system top-down from scratch will probably be beyond the abilities of human scientists, so a more likely approach will be to use an "evolutionary engineering" approach. The mapping between an artificial "DNA" sequence of molecular based growth instructions and the final 3D product, whether living or not, is probably impossible to predict due to its complexity, so probably the only method remaining is the one nature uses to learn how to "embryofacture" its creatures, namely evolution.

Evolutionary Engineering

An evolutionary engineering approach to embryofacture might work in the following way. One begins with a zillion random molecular "artificial DNA" strings, which translate themselves into blob-like 3D molecular structures. Predesigned molecular scale nanots then move in and measure how closely the actual blob resembles the shape or function of the microproduct that is desired. Those blobs that get a higher score will see their corresponding artificial DNAs survive and have more copies (children) made of them in the next generation.

The less functional blobs are killed off, Darwinian style, thus generating a kind of "survival of the fittest" strategy. The child-DNAs are then "mutated" slightly (i.e. the chemical instructions contained in the artificial DNA are modified somewhat). Occasionally, a mutated child-DNA will create a "fitter" (more performant) blob than its parent. By cycling through this loop many times, a desired artificial DNA is formed which grows the desired shape or function of its blob. The result is a molecular product that performs some useful function.

Self-Assembly

However, evolving single components is not enough. These components then need to be complementary in shape so that they

71

can "self assemble," i.e. fit together like jigsaw-puzzle pieces to form a greater functioning whole. Viruses form this way. Portions of DNA (genes) code for the construction of viral components. Once they are built, they click together to form whole viruses.

So the component parts need to have lock and key shape complementarities. They need to have the capacity to self assemble, simply by bumping into each other (as occurs frequently in the chaotic and high speed motion at the molecular scale).

This notion of self-assembly is very important when it comes to building an asteroid sized artilect, or even one of human size. A human sized artilect (or a human sized anything) contains trillions of trillions of molecules. To build a human sized artilect would require that all the atoms of that artilect, all trillions of trillions of them, be placed with atomic precision at just the right places. Such an artilect I believe would have to build itself through an embryological process. It would have to embryofacture itself. So how would such an artilect be built and designed in the first place?

Initially the first (very primitive) artilects would need to be built by evolutionary engineers (people like me). Perhaps they should be called "embryofacturers" or "embryological engineers." At first, the evolved 3D molecular structures could be assembled piece by piece into a larger 3D structure. Later, more sophisticated artilects could be built which perform their own evolution (perhaps within their own bodies) and make their own decisions at electronic speeds.

Of course, human beings would need to abandon all hope of fully understanding how these evolving, "Darwinian artilects" would develop. Their artilectual structure and functioning would be so complex and change so fast, that full human understanding of it all, would be totally impractical.

72

Such human ignorance will later prove to be powerful ideological fuel to the Terrans, who will argue that the very nature of artilect construction (i.e. Darwinian, self-assembling embryofacture) makes artilect behavior inherently unpredictable and hence potentially very dangerous for human beings. This point will be discussed again at length in Ch. 5, which presents the case of the Terrans.

Putting the Technologies Together

So let me try to summarize a bit here. After all, the point of this chapter has been to introduce those technologies that will enable the construction of artilects this century.

So far, the vision presented in this chapter is that of an artilect containing 10^{40} atoms or bits of information, using nanotech based, self-assembling, embryofactured, heatless, reversible, 3D, computer circuitry, thinking at least a million times faster (and probably a lot faster) than humans. It will contain a huge number of sensors attached to its surface, with enormous memory capacities etc. But there's more.

Quantum Computing Artilects

These artilects will be using molecular and atomic sized components, so these components will be subject to the laws of quantum mechanics. Recently the new field of "quantum computing" has become popular as theoretical and experimental physicists compete with each other to dream up new ways to "quantum compute" and to implement these ideas in real hardware.

I hesitate to describe what quantum computing is to the

general public. It is very counter-intuitive and difficult to grasp. If the following few paragraphs sound like gobbledygook to you, then just flip to the next topic. In a sense no one really understands quantum theory. It seems like a bunch of mathematical recipes that give good numerical answers to problems, but seems totally unintuitive conceptually.

Atoms behave in the weirdest ways, quite unlike what human beings are accustomed to at our scale of things. Quantum mechanics IS truly weird and abstract. It is a branch of mathematical physics that gives the probabilities of certain measurement results when atomic scale systems interact with human scale measurement devices. In classical mechanics, the state of a physical system is distinct, i.e. it has given values, e.g. its velocity at a given moment is V, its position is X, its kinetic energy is K, etc. In quantum mechanics, things are more abstract.

The state of a quantum system is represented by an abstract mathematical sum of numbers, where each number is associated with a measurement result if a measurement is performed. This summing and linear weighting of states is called a "superposition," and is the conceptual heart of quantum mechanics. Don't fret too much if you don't understand this. It is not essential to the understanding of this section.

It is this superposition that is the great feature of quantum computing. The superposition evolves over time, in a sense performing many calculations at once, whereas a classical computer can only do one thing at a time.

In classical computing, the state of a register (a storage chain of bits) is a definite string of 0s and 1s (e.g. 0011011101001). In a quantum computing register, the state is a superposition of a huge number of possible classical register states. For example, if there are N bits in the register, then there are 2^N possible different classical register states (e.g. if N = 3, there are 8 different classical states, 000, 001, 010, 011, 100, 101, 110, 111). If N is large, then 2^N is huge.

The enormous advantage of quantum computing is that this huge number of classical states gets treated as though it is just one (superimposed) state, one quantum state that the quantum system can handle. In order to perform a calculation with classical computing it is often necessary to test each classical register state one at a time, for all possible states. This is a slow business, and as N increases, the number of tests rises exponentially (i.e. like 2, 4, 8, 16, 32, 64 etc).

With quantum computing however, only one test needs to be done, because in a sense, all possible classical states are blended in together in the quantum register state. Quantum computing is potentially incredibly more efficient that classical computing. It is therefore not surprising that many physicists around the world are now racing each other to see who can build the next most performant quantum computer.

Since the artilect will be built with atomic scale components, it will need to function as a quantum computer. Since quantum computers are more efficient than classical computers, this is a good thing. The artilect will be a quantum computer.

The consequences of an artilect being a quantum computer are profound. Consider the 2^N times greater computing capacity of the quantum computer compared with the classical computer. An asteroid sized artilect would have 10^{40} atoms and hence bits. The potential computing capacity of such an artilect, even a classical computing type artilect, is hugely larger than that of a human being.

What then of a quantum computing artilect? If N is 10^{40}, what is 2^N? The mind boggles. When I say that an artilect could potentially have an artificial intelligence of trillions of trillions of trillions of times the human level, I am in fact exaggerating. My numbers are astronomically TOO SMALL.

Admittedly, state of the art quantum computers are handling about seven components. "Qubits" they are called, or "quantum bits." It has been possible to trap seven atoms in a row (an "atom register") in a kind of "magnetic bottle" and make them behave as a quantum computer to some extent. Quantum computing technology is still a long way from building a quantum computer with $N = 10^{40}$.

It is still debatable whether such large numbers will ever be possible, but recent "error correction" techniques etc seem to suggest theoretically, that quantum computers will become practical and later commercial. IF there are no theoretical reasons to suggest that such numbers are impossible, then that usually means that science will eventually find a way to create such machines.

Nanotech as a Brain Science Tool

Having said something (probably quite incomprehensible) about quantum computing, I turn now to another important question regarding artilect technology. The question is this. "How (human) brain-like will an artilect be?" My feeling is that as nanotech really comes on line and as Moore's Law really bites, our knowledge of how the human brain functions will increase dramatically. For example, over the past decade or so, various non-invasive techniques have been developed to observe the human brain in action without disturbing it in a fundamental way.

For example, mildly radioactive oxygen based fluids can be injected into the blood stream which travel to the brain and accumulate where brain cells are more active and need more oxygen and hence more blood. Human subjects are asked to perform various mental tasks, and as they do, the regions of the brain that are used more heavily in performing those tasks, show higher concentrations of radioactivity. Brain-scientists or neuro-scientists are thus able to localize where certain tasks are performed in the brain. Our knowledge of brain function, at least on a macro scale, has jumped considerably in recent years due to these new techniques.

More exotic methods employ phenomena based on nuclear physics, such as nuclear magnetic resonance (NMR) which I won't even attempt to describe here. Nuclear magnetic resonance imaging (NMRI), is getting more precise every year as new tricks are found to get finer spatial resolution and shorter measurement times of the human brain regions it observes. As Moore's Law provides faster electronics, the spatial and time window resolutions get finer and finer. Some scientists believe that by the year 2020 or thereabouts, it will be possible to observe individual synapses (neural connections) and their type, i.e. whether they stimulate or inhibit the firing of the neuron they connect to.

If this were possible, then neuro-engineers could simply "scan" the brain, i.e. just read it off, by downloading all the essential geographical information about the brain, e.g. the precise position of each neuron, and how and where it is connected to which other neurons, etc. All this information could be dumped into a "hyper-computer" that Moore's Law will make possible and then be subject to further analysis by other parts of the same computer. In a sense one would have a human like brain inside the hyper-computer.

This raises all kinds of moral issues. Can one switch off the hyper computer containing the human brain dump? Would that be murder? How do you define the essence of personhood? Does it depend on the technological base used to build that person, e.g. a carbon and DNA based technology, or a silicon based technology? If the essential architecture is the same -- if the functionality is the same, then should the silicon version of the person be given the same rights as the carbon based person? "Silicon rights?"

If humans decide that silicon based brain dumps are "non-people," then neuro-scientists and neuro-computer-scientists can start playing with the data (the person?). They will be able to make as many copies as they want of the data (cloning?). They will be able to perform all kinds of analytical tests on the data (vivisection?) trying to understand how the human brain functions. Our knowledge of how the brain works should then increase in leaps and bounds. As soon as the neuro-scientists provide new ideas on how the brain works, the neuro-engineers will be able to apply these new ideas to the creation of ever-smarter artilects.

Eventually, the neuro-engineers will be creating such powerful and intelligent machines, that the neuro-scientists will be able to test their hypotheses on how the human brain works by using the machines of the neuro-engineers.

At the moment, this marriage between neuro-scientists and neuro-engineers is pretty much a one way street, i.e. knowledge flows almost exclusively from the neuro-scientists to the neuro-engineers, but in time, as the neuro-engineers catch up, the information and idea flow will become increasingly a two way street. One day, the neuro-scientists will realize that the artilects being built by the neuro-engineers are in many respects outperforming the capacities of the human brain.

I will stop here for this chapter. There are many other examples of new and future technologies that could serve as the basis for artilect building this century. I hope that what I have presented so far has been enough to persuade you that this century's computer technologies will make artilect building possible, and that these artilects could have intelligence levels zillions of times greater than human beings. If I have also persuaded you that these artilects will be buildable before the beginning of the 22nd century, then this chapter has been successful.

But just because artilects can be built this century does not automatically mean that they should be built. (As the philosophers say, "Can does not imply ought"). The big question now is whether artilects should be built at all, and if so, what will the consequences for humanity be? What is likely to happen if the brain building Cosmists seriously intend to build artilects?

The remainder of this book attempts to answer this question.

This chapter has been rather technical and scientific in nature. The remainder of this book is more social, political, philosophical, ethical, even religious, and more appropriate to people who do not like to have to bother with scientific technicalities. For example, the next two chapters will present the views of the Cosmists (who are in favor of building artilects) and the Terrans (who are against building them).

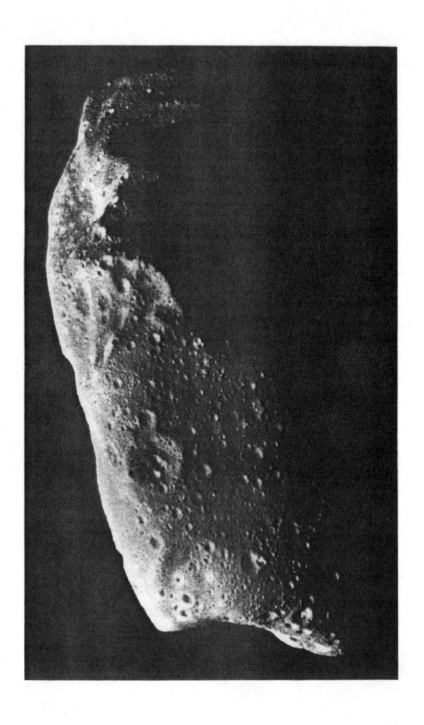

Chapter 4

The Cosmists

Before presenting in this chapter some of the reasons the Cosmists may use to justify building artilects, some initial remarks are necessary to clarify what is at stake in this debate between the Cosmists and the Terrans.

What is at stake is "the survival of the human species."

Never before has the stake been so high, and hence those arguments used by the Cosmists will need to be overwhelmingly powerful if they are to overcome the potential horror of the threat of the human species being wiped out. The enormity of the risk is so great, that I dare say most people, including myself, cannot even imagine at an emotional level just how large a potential tragedy we are talking about here.

In the 20th century, the Nazis wiped out 20 million Russians, the Japanese murdered 20 million Chinese, Stalin killed 30 million in his purges, and Mao starved 50 million Chinese peasants. These are amongst the greatest crimes in history, yet they pale in comparison to the size of the tragedy if ever the artilects decide to wipe out humanity. The tragedy would be total in the sense that there would no longer be any human beings left to mourn the disappearance of the species.

In face of this, the Cosmists, being human, must surely feel the incredible weight of the moral argument against them – "How can you even begin to think of taking the risk of seeing humanity wiped out? Cosmism is so monstrous, so horrendous, that you are inviting your own extermination at the hands of the Terrans."

All countries have jails to incarcerate murderers. Murder is not tolerated. The infamous mega-mass murderers of history, as referred to above, are truly hated by virtually all human beings. How then can the Cosmists seriously contemplate such a horror as the risk of "gigadeath," the annihilation of billions of human beings?

The answer to this question is found by appealing to counter arguments, which in the eyes of the Cosmists, are *even more powerful.*

As I write these words, I feel a shiver run up my spine. I have read too many history books which explain how new political or economic doctrines often begin with individual intellectuals or professors writing down their solitary thoughts, and how those ideas often end up generating wars that kill millions. Look at Rousseau's democratic ideas. Look at Marx's communist ideas.

This shiver going up my spine comes from the realization that the arguments in this chapter, the first of their kind, may one day serve as the intellectual basis for some future political movement, which in time I believe, will eventually result in the worst war in human history. This war will use the most destructive weaponry ever devised, based on late 21st century science and technology.

I feel terribly guilty in many ways, because I feel that my own work is part of the problem. As a professional brain builder, I am helping to create a technology that will enable artilects to be built. Initially they will be primitive, but they will keep improving. In the year 2000, I had a brain-building machine already built, capable of handling an artificial brain of nearly 100 million artificial neurons. I am hoping that the next generation brain-building machine that I

82

hope to build in the next few years will be a thousand times more performant than the first generation machine.

This day-to-day reality makes me very conscious of what is coming. Since I like to think that I am also quite well read in the humanities (politics, history, philosophy, etc.) I feel I have a greater insight into the longer-term social consequences of my work than many of my more engineering minded colleagues.

I feel profoundly schizophrenic about the work that I do. Deep down, I am a Cosmist. I often ask myself just how strong a Cosmist I am, and how much of me is Terran. I'm certainly not a 100% Cosmist. If I were, I would be quietly doing my work, not advertising its progress, and keeping dead quiet about the potential risks that massively intelligent machines may pose as a threat to humanity's longer term survival.

But I'm not a monster (I think). My second wife's mother was gassed by the Nazis at Auschwitz, so I know about mass extermination, about genocide, about mass horror. I lived with its consequences. When my second wife was a five-year-old Polish Jewish girl, she was handed over by her mother to a complete stranger on the platform of the railway station in Brussels, Belgium, where the cattle trucks were waiting to transport the Jews to the extermination camps.

To me, the Nazis were monsters. The Japanese were monsters, and are still unrepentant. Stalin was a monster. Mao was a monster. Am I a monster? Will my work as the pioneer of artificial brains, a technology and science that will very probably lead to the creation of massively intelligent machines within this century, inevitably lead to the creation of ideologies so murderously opposed, and with so much at stake, that a major war, the biggest in history, is almost inevitable?

If such a war does occur, killing billions, "gigadeath," doesn't that make me a monster, and the worst monster, worse than the monsters of Hitler, the Japanese, Stalin and Mao? Yet despite all this, I push on, because at the deepest level, I'm a Cosmist.

I think that NOT building the artilects would be an even greater tragedy. The more I think about the longer-term significance of artilect building, the more profoundly I feel I am a Cosmist.

But my mood swings. I will lie awake at night thinking rationally about the cosmic grandeur of Cosmism, about what these god-like artilects could do, and I feel the awe. Hours later I will wake up in a sweat, having been jolted out of a nightmare. I see in vivid scenes the deaths of my descendents in about a century or so, at the hands of the artilects, who have become so superior to humans that they see us as vermin.

The emotional reality and horror of it shake me. Normally I sleep rather soundly, so I don't remember many dreams, but this nightmare is recurrent, and so horrible in emotional terms that it is capable of waking me, despite my heavy sleep.

So, I feel schizophrenic about my work. I am profoundly torn, swaying between my head and my heart, so to speak. With my head, I think about the magnificence of the artilects, how godlike they could be, persuading myself with arguments that I will present in this chapter. With my heart, I am horrified at the concept and the prospect of "gigadeath," whether at the hands of the artilects directly, or as a consequence of a human Cosmist-Terran war whose primary causes I and other brain builders are now in the process of creating.

I had always thought that once the Cosmist-Terran debate had been well presented to the public that when push comes to shove, most people in their heart of hearts would be Terran. They would not want to see their families killed for the sake of some vague, abstract, emotionally distant goal of building the artilects, beings so distant and superior, that they will have almost nothing in common with us, so why risk paying such a terrible price for their creation.

I invite people at the end of my talks to vote. "Are you Cosmist or Terran?" I invite my readers to ask yourselves the same question once you have finished reading this book. The first time I asked an audience this question, the answer surprised me. I was expecting a 10% Cosmist, 90% Terran vote, but the reality was different. The vote split right down the middle, about 50/50. I thought that this might be simply due to the possibility that all these ideas were new to people, so they didn't really know what to think, and hence voted almost randomly. A random vote will almost always come out with a 50/50 result.

To try to change the percentages of the votes, I changed the content of my talks. I presented more clearly the horrors of "gigadeath," and showed more vividly the many powerful arguments of the Terrans, which I will present in the next chapter. But, the vote stayed near the middle, 60/40, 40/60, 50/50.

Gradually I began to realize, I think, that the vote was reflecting a deeper reality, namely that the Cosmist/Terran controversy divides people within themselves, i.e. within the individual. People would come up to me and say that they felt the same schizophrenia as I have been feeling for years. Only a minority of people were clearly in one camp.

Some individuals have come up to me and said that I deserved to be killed if I persisted. I would answer, "You are one of the first Terrans! In the future there may be millions even billions like you!"

But most people I find feel strongly both ways. I think the Cosmist case resonates emotionally with most people, as it does with me. I think it is in our human nature that the arguments I will present shortly in this chapter have a strong appeal. They are very strong arguments. They evoke our human sense of wonder, of exploration, of religious awe, etc. They are very strong arguments.

However, the Terran arguments are also very strong, as will be seen in the next chapter, so I just feel depressed when I think about what will come. With two powerful, bitterly opposed, and in fact murderously opposed, ideologies that are so evenly divisive, the likelihood of a major conflict is only increased.

The bitterest of wars are often civil wars, where brother kills brother. The US civil war, fought largely over the issues of the morality of slavery and the right to secede from a union of states, still divides ideologically, to some degree, the northern from the southern US states.

The Terran/Cosmist conflict will be a kind of planetary civil war, because it will probably not be correlated with geography, i.e. with one geographically localized group taking up more the Terran case, and another geographical area taking up more the Cosmist case, although this possibility is not excluded. Well into the 21st century, the global telecom capabilities will probably be powerful enough to keep most of the world's citizens well informed about the conflicting ideas in this great debate.

People will make up their minds according to their own personalities, abilities and interests, and not be "brainwashed" so much by their local media. The media will be largely global by then anyway, with thousands of sources of news and ideas coming from all over the planet.

Having now provided above a kind of lead into the arguments used by Cosmists, I will now begin a more systematic presentation, argument by argument, in an order starting in my opinion from the most important and most powerful. I give each of these arguments a name.

I am expecting that these arguments are only the first few of many that other "Cosmist intellectuals" will invent in the future. Since I believe the Cosmist/Terran ideological and later military conflicts will dominate our century, it is only reasonable to expect that many first class minds will apply themselves to the intellectual rationalization of the two rival cases.

As I start to write down these arguments, which I have been thinking about for years, another shiver goes up my spine. I feel a heavy responsibility for the future quality of life of billions of human beings. I feel in some ways that I may have started something that will eventually destroy them, and that in one sense I am doing this by choice. In another sense I feel I do not have a choice, because the force of the Cosmists arguments is so strong that I feel that I don't have any real alternative. The Cosmist case for me is just too compelling.

Arguments in Favor of Cosmism

1.   The "Big Picture" Argument

The strongest argument in favor of Cosmism in my own mind is the one I label the "Big Picture" argument. It has to do with the feeling that human existence is so petty, so trivial, so banal, so insignificant, that there are bigger things in life than those that concern human beings on a daily basis.

Science teaches both me and anyone who is interested to learn, that we humans live on a planet that orbits a very ordinary star, that is one of 200 billion in our spiral galaxy. Our galaxy in turn is only one of billions in the visible universe. Furthermore, according to modern cosmological theories, there are probably zillions of other universes. In other words, our petty little human lives, which last an ephemeral three-quarters of a century, are utterly negligible in comparison with the age of the universe, which is billions of years old. As humans, we are nothing. We are of zero significance on the cosmic scale.

Modern science has discovered that the laws of physics and chemistry are the same throughout the universe, so it is almost certain that "out there" a zillion different biological civilizations have evolved, that they have reached a stage of technological intelligence and then built artilects to supercede them. It is therefore quite possible that there are zillions of artilectual civilizations in the universe, at all kinds of different stages of development. As humans, we are probably too stupid to be aware of their presence, and are totally ignored by them, due to our extreme primitiveness. To such artilectual civilizations, billions of years older than we are, we would seem to them as a primitive moss would seem to us, that is, totally uninteresting.

This would answer "Fermi's question." Fermi, the famous Italian/American nuclear physicist, who played a major role in developing the nuclear bomb, was cynical about the existence of ETs (Extra Terrestrials). He asked that if they exist, then "Where are they?" We have no sign of the existence of ETs, no proof. If life is common throughout the galaxy, as physics and chemistry suggest it should be, then where is all this life manifesting itself? It may be all around us, but we may be too stupid to recognize it.

To get a feel for the "big picture," you only have to look up at the stars. Alternatively, you can stare at a glossy color astronomy book, with large color plates of galactic arms showing millions of stars, where each little white dot in the photo is a star, a sun like ours in many cases, which probably contains a solar system. Their huge number is humbling. Our sun is as significant or as insignificant as any one of those zillions of stars.

From the galactic point of view, would it matter much if the human race were wiped out? I think the universe wouldn't give a damn. This kind of thing may have been going on for several billion years, as biological civilization after biological civilization reaches intelligence and then destroys itself either directly by its own doing, with its own weapons, or indirectly, at the hands of its artilectual creations.

The "big picture" argument is admittedly an intellectual one, and may not mean much to most people. The majority of folks appear totally indifferent to the fact that there is a much bigger existence "out there." It would therefore not surprise me that as the artilect debate heats up, we may see a correlation growing between being Cosmist and being more intelligent. I suspect that those people with higher IQs are more likely to be open to the Cosmist perspective because they will be better educated and aware of the cosmic realities that science teaches.

When I was younger and traveling around internationally on a shoestring budget, staying at youth hostels, I was always struck by the high percentage of young hostellers who were university graduates. They were not at all the usual cross section of humanity. The higher intelligence seemed to translate into a higher degree of curiosity about other cultures' lifestyles and values. I suspect the

Cosmist perspective will appear more to the more open-minded and more thoughtful person. I may be wrong on this, but that is my guess. (However, you may doubt this due to the 50/50 vote results I mentioned earlier. But most of my audiences were at universities. I will be curious to see how the general population will vote in future decades.)

To the Cosmist, building artilects would mean that they could become part of that "big picture," whereas as human beings, we could not. For a start, we are too ephemeral. We die too quickly. An artilect could be made to be immortal, and hence have all the time it needs to do whatever it wants. As humans, we are too stupid to figure out how to escape easily from the prison we call Earth. Our bodies and minds are too primitive to take a form that would allow us to be more cosmic creatures, to voyage easily outside the cradle we call Earth.

If we consider how much scientific progress we have made as human beings in the past century, with our puny human brains, consider what an artilect could do with its giant brain and over billions of years. It would be so much more capable of discovering the secrets of the functioning of the universe and could use those discoveries to empower itself. It could use phenomena that we as human beings don't even know exist, or could not even understand if some artilect tried to explain them to us.

The more one reflects on such things, the greater the sense of awe one feels. I see this as a kind of "religious" feeling, similar to the religious longings of earlier centuries before the rise of modern science.

Since I am now touching on religion, this brings me to the second important argument in favor of Cosmism. It sees Cosmism as a "scientific religion."

2. The "Scientific Religion" Argument

I am not at all religious in the traditional sense. I look upon traditional religions as prescientific superstitions. They are quite incompatible with what modern science teaches us about the way the world is. My level of scientific learning and my critical cynical scientific mind cannot accept such doctrines, which in my opinion cannot hold up to critical scientific and analytical scrutiny.

Yet, as I get older, (I'm now in my late 50s as I write this), I find myself getting increasingly interested in religion, but not because I find the traditional beliefs any less incredible, but because I feel I have a greater understanding of the religious impulse. I find myself thinking more about the deeper aspects of human existence, about the bigger things besides the day-to-day distractions that seem increasingly petty to me. Probably the greatest motivator for thinking about such things is the brute fact that so many of my friends are dying.

I have quite a lot of friends who are older than I am, in their 60s. Since most of them are male, and given the average life expectancy for males in the first world countries is about 75 now, that means it is statistically expected that a growing, non negligible proportion of them will be dying off in their 60s.

My second wife died in 2000. Admittedly her premature death was largely her own fault, and partly the fault of the cigarette companies. She smoked heavily before she met me, and died of lung cancer at the early age of 62, young for a woman.

There is nothing like the death of people close to you, to make you think of your own mortality. The older I get, the more I reflect on it, and read a lot about the biology of death, about why cells age, about the technology of immortality, to get a deeper insight into why we are so terribly mortal.

Religious belief is a cultural anthropological universal, one of the very few, along with for example, the taboo on incest. The need for religion is very strong, as evidenced by the fact that it is so ubiquitous.

My scientific curiosity and scientific knowledge prevent me from being traditionally religious, but the initial craving for some deeper "spiritual" understanding of human existence and mortality, I do feel. I suspect that the scientific hunger I have had since a teenager has a lot in common with what drives a lot of very bright men into theology.

I am a born scientist. I like to read and think a lot about what I call "the big scientific issues." For example -- evolution and the billions of years it has taken to generate life; about astronomy and its huge time frames and distance scales that dwarf human affairs into insignificance; about theoretical physics and its conceptual mysteries, "How can nature be like that?"; about embryology and the intricacies of building a body from a single egg; about brain science and the mystery of how thought and consciousness arise from the interaction of billions of neurons, and so on.

I like to go to good science fiction movies or to read hard science fiction books, especially those written by professional scientists, hoping to experience a sense of awe, of magnificence, of the "big picture" of things.

As I said earlier, I have many books on astronomy, particularly those with large glossy photos of galaxies with their zillions of stars. I love to ponder them, staring at them, wishing that today's imaging technology were more advanced, so that I could experience the emotional rush of looking at a 3D holographic image of deep space, of a spiral galaxy filling my living room with an image as authentic as if I were in deep space myself.

This technology is coming, and when it does, I believe it will create a much deeper sense in the global public mind of what I call "space consciousness."

By "space consciousness," I mean the awareness of how huge our universe is, and how insignificant our human lives are in comparison. It would create a sense of religious awe in most of us I believe.

This new century's media technology will give humanity the wonderful, awe-inspiring, 3D images I mentioned above. I believe such images will have more impact on the human psyche than did the first photos of the terrestrial globe that were sent back from the moon missions in the 1960s. Since we don't have such realistic 3D images yet, it is difficult to imagine the emotions they will evoke. I get a taste of what it will probably feel like, from watching certain science fiction movies on a large screen (2D of course), when one sees the stars in the background.

If you live far from the smoggy cities and away from street lighting (and not many people do nowadays), or you are traveling on a cruise ship in mid ocean, you will be able to look up at the stars. You will not see spiral galaxies in all their incredible beauty -- you will need a telescope for that, but you will see thousands of twinkling stars. That image alone should make you feel insignificant if you know any science -- i.e. that each of those twinkling white dots up there is a sun, probably with its own planetary system. It's now thought that most stars have a planetary system and that they have been twinkling away for billions of years in most cases. It's a humbling experience, is it not?

Now imagine that you are living a few years into the future

and you switch on your 3D-entertainment system and choose by voice command to your holo-viewer, a 3D image, and room size, of a spiral galaxy. A magnificent, breathtaking, awe inspiring, emotionally engulfing image appears, together with powerful hypnotic music, creating a totally enthralling experience, especially if it's the first time you have seen such a thing. You ask the viewer to zoom in at the bottom right corner, and the disk appears closer. You ask for continuous zoom.

Gradually the image changes from a foggy mist into zillions of white and multicolored dots of light. You know these are stars and nebulae. The zoom continues and the dots become more distinct, and some of them grow slightly larger. One particular dot now grows into a small circle as the background mist crystallizes into a spectacular image of millions of clear dots of light in a vast spiral arc.

You ask the viewer to zoom in on the growing circle. Some moments later, the size of the circle grows until the room is filled with its light. It is too bright. You ask for less light. You hope that this star has planets.

You ask the viewer to see if it has any. The image shifts to the right and stops. You see nothing. You ask for a zoom and a brightening. You see a small sphere. It is a gas giant planet. You zoom in. The planet fills the room, but all you see are multicolored clouds of gas, no surface. You ask the viewer to look for an inner planet. It finds one, a blue sphere similar to the Earth. You get excited.

You zoom in and notice the oceans and landmasses. It is the Earth. You are disappointed. You continue the zoom so that a particular land mass grows and grows until you can make out a city, then a city block, a park, a young couple having a picnic in that park, then the palm of the man's hand, then his skin, his skin cells, his blood cells, his DNA, then molecules, atoms, a nucleus, a nucleon, a quark, and then the image freezes. You have reached the limits of humanity's knowledge.

What I have just imagined is not original. A book containing such images appeared some years ago, called "Powers of Ten." It begins with an image of the whole universe, where the galaxies are just tiny dots, right the way through to quarks, where each image, accompanied by an explanatory text, is a zoom-in ten times closer than the previous image.

The first time I saw this, it fascinated me. It made me realize how incredibly limited our daily human experiences are when compared to what there is to be seen and understood "out there" at the macro-scale of the universe, and "in there" at the micro-scale of the atomic world. Such experiences make me think how limited in scope our normal human lives are. We live in the middle range of these size scales with roughly as many orders of magnitude that are larger than us as those that are smaller than us.

Our human brains have evolved to be preoccupied by those phenomena of similar scales, both in terms of space and time. As humans we have such a limited view of the world, such tunnel vision, metaphorically speaking.

But thanks to our modern scientific knowledge, we know at least that such scales exist. Most of these scales were only discovered within the last hundred years or less. This knowledge generates in me a fascination for science. I find myself caring more and more about what I called above, the bigger things, than the humdrum of daily existence, which is snuffed out in 80 years or so anyway, due to our evolutionary programming, our programmed death. As I get older, my mortality weighs more and more heavily on my consciousness, making me think that it would be nice to live longer, to be immortal maybe.

Humanity could give itself immortality if it wanted to. Mortality is just a molecular program written into our DNA. The DNA in our reproductive cells is immortal and can elaborately self repair over the generations. It recombines with the DNA of another person during sex, and creates a whole, fresh, perfect, new body. Bacteria are immortal. Cancer cells are immortal. It can be done. It would mean merely changing the molecular programming in our cells. Life is a program, and so is death. Death can be unprogrammed. Such unprogramming we can have -- and probably within a century, as humanity understands more about the molecular biology of how death occurs.

If I had the power and the knowledge to make myself immortal, then why not, while I am at it, increase my intelligence as well. The intelligence level I have already, allows me to experience the joy of understanding how mysterious and wonderful nature can be at the level of the atomic world. For example, the study of quantum mechanics has always been a joy to me and was the topic of my first research efforts.

If I could just increase my intelligence by 10% or so, maybe I could appreciate with less intellectual effort the great beauty in modern "string theory" that the best of modern theoretical physicists talk so much about lately. If I could increase my intelligence by a further 10%, perhaps I could feel the awe that Einstein said he felt at being able to discover some of the deepest of mathematical relations describing the behavior of the universe. The fact that humanity finds nature comprehensible was always something that he found to be incomprehensible. It truly baffled him.

If I continue thinking along these lines, why limit myself to human levels of intelligence. If I can attain god like powers and make myself immortal and more intelligent, then why not a lot more intelligent, like tens times more, or a million times more, or trillions of times. Why not just turn myself into a god or an artilect?

I then start thinking that if I become an artilect, with 10^{30} components or more, what would be the meaning of "I?" The original human "I" would be totally swamped by the vastly greater mass of computational, memorial and sensorial components. "I" would no longer be "me." "I" would be a new "me," an artilect. So why bother going through all the transitional stages from human to artilect. Why not just build an artilect directly from scratch, and treat it as though it were my own child, i.e. not me, but my offspring, the creature I created? In a sense if I become an artilect myself and I make this transition quickly, then my old self would die in a sense. My old self would be drowned in and totally ignored by the vastly superior capacities of my new artilectual self.

As an artilect (irrespective of whether I started off as a human being or not) I could do my own "powers of ten" travel to a large extent. I could attach all kinds of scientific instruments to my body, integrating them as part of me and observe what I wanted. Assuming that the limit of the speed of light remains in force, I would not be able to travel too far too quickly, but at least I could observe the very small size scales with ease. But, with greater intelligence, perhaps I could discover ways around the speed of light barrier, perhaps by using space-time wormholes and the like, or perhaps by using phenomena that human beings, with the intellectual limitations of human beings, cannot even imagine.

It would be nice to be an artilect, a god, a supremely powerful

97

omnipotent being. I could be such a creature by late 21st century and beyond. It's possible. It's not an unattainable dream.

I am not a poet nor a playwright, so my attempts here to convey the sense of religious awe at becoming an artilect will need to be expressed in a more emotional and convincing way by real professionals of the arts. (I encourage artists to create such "artilectual" works.) All I can do here is attempt to convey some measure of the strength of "religious" feeling that I and other Cosmists will make public this century.

Cosmism to me is a kind of religion, one compatible with scientific knowledge, and hence acceptable to my critical scientific mind. It's a "scientist's religion," but you don't have to be a scientist to have the same feelings of religious awe when contemplating the potential of what artilects could be. I may be a scientist, but I am also a human being, and hence feel the same religious pull as nearly everyone feels at some stage in their lives, when faced with the deepest of questions.

There is something truly magnificent about the Cosmist goal of building artilects. The artilect itself is godlike. Building artilects would partly satisfy in me some deep spiritual quest that I have difficulty defining clearly, even to myself. The artilect is a kind of god to me, of a type that I can believe in, having immense power, and yet one that I and others may help build in the future. That would make me feel powerful, but its more than just a question of power. It's also a feeling of wanting to go beyond, way beyond, our current human limitations. I will write more about this shortly. It is also about wanting to leave a mark after I'm dead. It was this desire, I believe, that motivated the pharaohs to build the pyramids. An artilect would be a magnificent pyramid.

Some males become monks in order to contemplate a higher

form of existence. I understand the impulse to do this, because I share it in many ways, but I cannot go along with the beliefs and life styles of traditional monks and nuns, whose lives I feel are wasted in the pursuit of beliefs that to me are pure fictions. To my mind, they lead sexually, emotionally, reproductively, and existentially impoverished lives.

I cannot take traditional religions seriously, since they are incompatible with what I have learned about the world. But the impulse to be religious is there, and goes unsatisfied. So it probably will not surprise you to learn that the day that the idea of "Cosmism as a Scientific Religion" occurred to me, I was deeply moved. Cosmism as a "religion" would satisfy a lot of my "spiritual" needs, and importantly, would be compatible with my scientific worldview. It would be a "scientist's religion" and that for me and perhaps for millions of people, may prove to be one of Cosmism's strongest attractions. If a lot of scientists feel the same way as I do in the future, then it is more likely that the artilects will be built.

But what exactly is the source of this attraction? If the Cosmists are prepared to risk the start of an Artilect War and even the extermination of the human species, due to the strength of their desire to build artilects, no matter what, just what is it that motivates them so powerfully? What is so godlike about the artilects that makes Cosmists so committed to building them?

3. The "Building Artilect Gods" Argument

Another very powerful argument the Cosmists will employ is the sheer attractiveness of the prospect of building godlike artilects. To many Cosmists, this attraction will be compulsive, overriding all others, and motivating any means to achieve this glorious goal, even if the human species has to risk being wiped out as a result.

Let me try to convey more clearly and in more detail the godlike qualities that artilects could have. (For further discussion along this line, see Ch.7 on the Artilect Era). If you find yourself mesmerized by what follows, perhaps you will be a lot more sympathetic to the dreams and obsessions of the Cosmists.

An advanced artilect could be very large, e.g. the size of an asteroid. If it were of planet size it could orbit about a star and absorb its energy. If it were in the shape of a huge hollow sphere with the star at its center (a "Dyson sphere"), it could absorb all of the radiated energy of that star. If such an artilect is built in our solar system, the material necessary for its construction could be taken from the asteroids in the asteroid belt, perhaps all of them.

So potentially, such a creature could consist of 10^{40} or even 10^{50} atoms, and hence bits. The molecular or atomic size switching elements would be switching (flipping from 0s to 1s or vice versa, which is a fundamental operation in computing) in femto-seconds (a thousandth of a trillionth of a second), so altogether, the artilect could be switching at about 10^{55} or 10^{65} bits a second. This is an astronomically large number.

Compare this with the equivalent switching rate of the human brain. The information processing of the human brain occurs (arguably) at the synapses (the inter-neural connections) at a rate of about 10 bits a second. Since there are about 10^{15} synapses in the human brain, that means the total brain processing speed is about 10^{16} bit flips a second.

The artilect's processing speed is thus 10^{40} or 10^{50} times greater, which is trillions of trillions of trillions times more. Such numbers are so large, that it's difficult for human beings to absorb their significance. Let me try to spell it out a bit more clearly.

Such creatures would be capable of "living the lives" of zillions of human beings in a mere second of their existence. A human life of about 80 years (80*365*24*60*60 seconds) i.e. 2.5 billion seconds, computing at 10^{16} bits a second, over an average human life time would process 10^{25} bit flips total. So an asteroid sized artilect, for example, with 10^{40} atoms, could process the equivalent of 10^{30} human lives per second, i.e. a million trillion trillion lives. That's more than the number of atoms in an automobile.

But sheer processing speed is only the beginning. What is truly significant and godlike about an artilect would be its ability to use that speed in fascinating ways. For example, it could absorb matter into itself from the asteroids and reassemble it into computing material to do whatever task it sets itself. In fact, the above talk of an artilect performing a single task at a time, is probably a joke. An artilect would probably be thinking a zillion thoughts at the same time. It has the luxury to do so, because it has enough matter and speed to allow it to do so.

How would the artilect know how to arrange the matter to think the thoughts it wants? Well, it could employ Darwinian evolutionary experiments on parts of itself and examine the results with other parts of itself. The newly and successfully evolved parts could then be absorbed into its general structure. These experiments could be going on all the time, and at incredible speeds. The intelligence level of the artilect could be increasing astronomically every second.

The artilect could have a huge number of sensors on its

surface or interior. It could build artificial life forms and play with them as part of itself, learning about life processes. (Maybe some super artilect is doing this right now with our universe -- more on this in Ch.7).

The artilect would have the means to amplify continuously its intelligence to levels human beings cannot imagine. If intelligence is correlated with processing speed, memory capacity, etc. then obviously the artilect could be trillions of trillions of trillions of times more intelligent than human beings. For example, imagine it takes an artilect a trillion atoms per computational module to perform some basic task. How many such modules could it have? Trillions of trillions and more.

Mouse brains cannot perform certain functions that human brains can, because they don't have enough brain modules of appropriate structure. An artilect's modules could be evolved and deployed to perform zillions of functions, while at the same time evolving and restructuring itself.

The artilect would be the consummate scientist. It could manipulate and examine its own matter. It could transform elements (e.g. from oxygen to carbon) from its own body, or just select and use appropriate atoms from its own storehouse, which would also be part of its body. Alternatively, it could convert parts of itself into transporters and fetch material from elsewhere. With the full range of chemical elements (from hydrogen to uranium and more) at its disposal, it could design and build its own experiments to investigate its own structures. The knowledge it would obtain it could use to redesign itself in better ways. The artilect would learn zillions of times more about the world and itself than human scientists will ever know. It would be truly godlike in its knowledge and power to manipulate the world.

As its knowledge of the world increases, i.e. its science, it could transform itself via that new knowledge. It could apply technological principles to itself, the way human scientists and technologists tend to do with objects external to themselves.

Human knowledge is said to double every ten years or so. Let us call the total quantity of human knowledge at the year 2000 a THKU (Total Human Knowledge Unit). What would the artilects rate of knowledge growth be in THKUs per second? It takes 10 years for roughly ten billion people to double their knowledge. Even if the artilect had the same intelligence level as humans per unit of matter (which we say above is unlikely) it could still vastly outperform the human population because of its much larger mass and processing speed.

10 billion humans processing for 10 years is how many bit flips total? That's 10^{10} (the number of people) times $10*365*24*60*60$ (the number of seconds in ten years) times 10^{16} (the bits-per-second processing speed on one human brain), that is $10^{(10+9+16)}$, i.e. 10^{35} bits. An artilect can process 10^{55} bits a second, i.e. 10^{20} THKUs per second, i.e. nearly trillions of trillions. Of course with its vastly superior intelligence level it could do the above zillions of times faster, but that would be hard to calculate.

If you don't follow the math, don't worry, just accept the bottom line that the artilect is doing everything faster and better than humans by factors of trillions of trillions at least.

I should add that the above calculations are based on traditional "classical computing" principles. If such an artilect were to use "quantum computing," the resulting numbers involved would make the above numbers seem hugely too small.

Probably books will be written shortly on the potential capacities of artilects, and I hope this book will inspire such authors. I could probably go on and on about how astronomically superior an artilect could be.

I hope the above is enough to show you that the artilect is a truly godlike creature, so vastly above human capacities that it is an object of worship to someone like me who builds brains for a living. If you were a brain builder, the artilect vision would be like a great shining beacon beckoning you on with hypnotic force. It would create a strong sense of religious awe, and best of all, it is entirely compatible with science, thus making it much worthier of "worship" than traditional beliefs.

You don't have to be a scientist to appreciate this. Scientists may be able to savor the vision more easily because of their abilities and knowledge, but to anyone who enjoys thinking about such marvels, the artilect vision I can imagine could be truly enthralling.

Not only is the artilect something compatible with science and something worthy of devoting one's spiritual energies to, but more importantly, it is real, in the sense that it is achievable. It is buildable. Creating such creatures would be possible, if human beings wanted to do this. Human beings, the Cosmists, could become "god builders."

I believe that building such creatures, or at least their early precursors, will become the life goals of the Cosmists. It is a magnificent dream, truly awe inspiring, mind stretching, energizing, life orienting, meaning giving -- in short, it is a "religion."

Look at the way the Arabs were suddenly energized by the (human) invention of Islam. Suddenly millions of Arabs had a set of beliefs that galvanized them, gave them a sense of purpose, excited them, and channeled their collective energies to conquer all of north Africa in the 7th century and even into Spain. If it weren't for the Arabs and their connections with the ancient Greek and Roman world, Europe would not have had its renaissance via Spain.

Look at the Christians and their spread over the western world. Look at Buddhism and its spread in the east. All these belief systems direct human lives. They contain ideas that people, billions of people, devote the energies of their lives to. A similar situation could arise with Cosmism. The Cosmists could devote their lives to the achievement of building the artilects. I am a Cosmist. I build artificial brains, although obviously nothing like what I have described above. Nevertheless it's a step in that direction. Building advanced artilects is a long-term dream of mine -- not one I will see in my lifetime, but I can hope to be the dream's prophet. I can hope to inspire future generations to adopt that dream.

I believe that the Cosmist vision will give humanity a new religion, a very powerful one, suitable for our new century and beyond. Like most powerful religions, it will generate energy and fanaticism, as people channel the frustrations of their daily lives into opposing those people who oppose their own beliefs. In this case the opposition will be the Terrans. Major religions have created major wars in the past. Look at the crusades between the Christians and the Moslems in the Middle East, or the Catholics and the Protestants in Europe.

I believe that this new religion will also help create the Artilect War. The fanaticism and strength of purpose generated by the Cosmist vision will be pitted against the fear of the Terrans, two extremely powerful forces. The war will be passionate and very deadly, given the historical era in which it will take place, i.e.

probably late this century with late 21st century weapons.

My intuition tells me that the above "religious argument" is probably one of the strongest that the Cosmists will possess. The vision of what an artilect could become will be so powerful that even if the Terrans totally exterminate the Cosmists, the vision will always be there to inspire new generations of Cosmists. It will not go away. A powerful new idea, whose time has come or soon will come, can move mountains, planets, even universes.

4. The "Human Striving" Argument

After the preaching of the above argument, I turn now to another that I believe the Cosmists will use to justify the creation of artilects. I call it the "human striving" argument. Why do human beings always seem to want to go beyond what is currently known, currently explored, to climb higher peaks, run faster, cure more terrible diseases, become stronger, fitter, more brilliant, and excel at one's work? Why this constant pushing at the barriers? I believe it's built into our genes. Evolution has made us this way.

Chimpanzees show a strongly developed sense of curiosity. Human beings, and especially children and scientists (big children) have an even stronger sense. Our big brains evolved to discover how our environment works. If we have a better knowledge of the dangers and delights of the world that surrounds us, then we are more likely to survive. But if we lack a curiosity to explore our world, we learn about it more slowly.

Those apes and humans who learned faster by being driven to explore, to push the limits of the known, learned faster and hence were more likely to survive. Well, not always. Some poor chump

106

had to be the first to discover that arsenic was poisonous, but his neighbors learned from his death. Since they all had the same curiosity/striving genes, they learned from his negative experience.

Is it not inevitable that once the prospect of building artilects is with us this century, that our genetically determined striving curiosities will propel us towards building them? Can we help ourselves? Will we have to build them the way Hillary had to climb Mount Everest, simply because the challenge presented itself, and the technology and management techniques had developed enough to make the mountain conquerable?

Look at how humanity has explored the continents. Early man left Africa in search of food and fresh territories, roaming across all continents, building boats to travel long distances guided by the stars. As technology improved, ocean faring ships were built which allowed sailors to discover new worlds. In the 20th century, humanity began to explore space. We have even set foot on the moon and will soon set foot on Mars.

The will to strive and explore may also be motivated by boredom. Consider the following scenario. The world economy is growing by several percentage points a year on average. Thanks to compound interest, this means that the economic welfare of nations increases at an exponential rate. Already, several first-world nations live in real affluence, in the sense that they are well fed (if not over fed), kept healthy, are well-educated, amused, and live long lives. It is only a question of time I believe before the whole planet will become affluent in this sense.

I believe as telecommunications improve, for example as digital satellite TV beams thousands of international channels from the sky, a snowball effect towards having English, as the world language will be reinforced. Eventually, probably everyone on the planet will speak it at least as a second language. Ideas will then travel rapidly, resulting eventually in a largely homogenized culture, perhaps not totally, but generating enough trust for the creation of a world government. Then the enormous funds wasted on armaments can be spent on improving the standard of living of the world's citizens.

And then what? I predict that a global ennui will set in. Humanity will need a major new goal to challenge itself. What better goal than aiming to build artilects. It is a goal worthy of the level of human skills, as they will exist later this century and beyond. It will be a truly global goal, affecting everyone on the planet. It is doable this century rather than later, and hence the timing will be good. The ennui will be felt very strongly this century, because the world will reach affluence this century, even in Africa, the poorest and most backward continent.

If the Terrans win, and humanity decides not to build artilects, I can imagine a lot of bored and frustrated people twiddling their thumbs, just itching to climb up the evolutionary ladder, so to speak. Personally I think it will be almost impossible not to go Cosmist. It is in our human nature to strive, to be curious, to go where no man has gone before.

5. The "Economic Momentum" Argument

The next two arguments are not what you might call "active arguments" that Cosmists would need to give their intellectual energy to. They are more passive arguments, in the sense that they will be influential almost by default, independently of how much energy the Cosmist intellectuals give to pushing the above active arguments.

The next argument in favor of Cosmism, I call the "economic momentum" argument. I believe that such a powerful economic and political momentum in favor of Cosmism will be built up over the next half-century or so, that stopping it will be almost impossible.

The advanced artilects will be the offspring of earlier simpler artilects, which in turn will be the offspring of the artificially intelligent, artificial brains that people like myself and others will be developing early this century.

Consider for a moment some of the massively successful AI products that we can expect to see developed in the next few decades.

We are beginning already to talk with our computers. As the years go by, these machines will become conversational computers. Call them "talkies." Since a lot of people live alone and get lonely, there will be a huge market for such machines, which will get smarter, more emotional, have a richer vocabulary, with a greater learning ability, larger memories etc, over the years. In time, people will start having better "relationships" on a conversational basis with their talkies than with other people. These conversational computers will be able to adapt to their human owners by building up a knowledge base of their owners' interests, intelligence and knowledge levels, and behave towards their owners in as familiar a way as a spouse does after many years of marriage.

Of course, such a high level of technical sophistication will not come overnight. I believe however that there will be such a high demand for such products, that they will eventually be built, and that it will be the steady increase in their artificial intelligence levels that will alarm the Terrans. I will discuss the Terran viewpoint in the next chapter.

In time, vast talkie research and development industries will be created to satisfy the enormous demand. Social intercourse is a deep need, and as the talkies get better at it, demand from the public will grow.

A similar story will occur with household robots. At first, they will perform only very simple tasks, such as vacuuming the carpets, and sweeping the floors, but as artificial brain building develops, the number of tasks these "homebots" can perform will increase.

They will be given the ability to understand the human voice, so they can obey commands spoken by their human owners. Perhaps they will be made into talkies as well, so that they can talk back, giving explanations. "Why didn't you sweep the floor today?" "Because you forgot to replace my battery this morning."

As the years go by, homebots will become increasingly useful. Huge R&D efforts will be invested into them and they will be sold to virtually every household. They will become the "big ticket" consumption items of households, as is a car today.

Another class of AI products that we can expect will be teaching machines, "teacherbots." These machines will adapt to the intelligence, knowledge, interest and curiosity levels of individual users. Human students will be able to learn at their own individual rates, instead of the incredibly clumsy schooling methods we use now. In today's schools, a single human teacher attempts to educate a few dozen students simultaneously, pitching the intellectual level of the presentation at the middle ability range, thus leaving the intellectually slow behind, and leaving the bright bored.

Teacherbots on the other hand will be able to educate students individually. They will become far more efficient than a human teacher, presenting material in a way that fascinates. A human student whose curiosity is aroused can learn avidly and long. The intellectual accomplishment level of the whole society should thus rise considerably.

The teacherbots will tap into knowledge bases around the world, hunting out information relevant to the needs of their individual students. They will in effect become sources of infinite knowledge and fascination to those who really want to learn about some topic in detail. Of course, such educational facilities will also rapidly expand the knowledge gap between those people who will be motivated and hungry to learn and those who will not care, but globally speaking, the general level of awareness and absorption of knowledge will increase dramatically.

Teacherbots, along with the above talkies, homebots and other such products, such as sex robots, baby sitter robots, etc, will generate a huge industry. These are examples of how AI based products will probably form the foundations of an AI based world economy.

Very powerful, strong egoed individuals will manage the creation and expansion of these trillion dollar industries, investing large amounts of money into their research and development. Over the years, millions of people will be involved not only in using these products, which will be universal, but also in researching them, designing them, and building them. AI based products will form the skeleton of the world economy. They will form the basic industries of the early decades of this century, the way the automobile, oil, insurance, etc. industries, did in the 20th.

111

Once millions of people's livelihoods are tied up in the creation and use of artificial brain based products, how will it be possible to stop the development of the AI based economy if ever the Terrans decide such a thing is necessary? Increasingly, the big egoed powerful men of industry and politics will begin to use their powerful minds and their influence to push their own agendas onto everyone. That is the nature of power. This is nothing new. Powerful men have had their way for thousands if not millions of years.

In the next chapter I will give the Terran point of view, but I need to anticipate a bit of the discussion in that chapter here.

How will the industrial magnates of the brain based computer industries react to a growing Terran fear of the rising intellectual powers of the early artilects? These magnates will have devoted their whole lives, their egos, their very souls, to artilect creation. As leaders of their industries, they will have selected themselves as the most capable people, the most visionary, the most forceful, the best organized, to drive their industries forward. Such powerful men will not give up easily their life's work to appease the fears of the Terrans, although it is possible that they might become Terrans themselves, as a result of their experiences.

However, in most cases, I consider it likely that the leaders of the artificial brain based industries will prove to be powerful Cosmists, because it will be very much in their self-interest to be so. To minimize the fears of the Terrans, these captains of industry will try to make their products as human friendly as possible. They will make them "warm and fuzzy," so that they will appeal to human nature.

But there is a limit to the extent to which the growing computational power of their products can be hidden. The sheer computational miracles that these early artilects will be able to perform will be increasingly obvious, no matter how warm and fuzzy their packaging. Sooner or later, millions of people will become conscious how fast and how smart these earlier artilects are becoming. The "artilect debate" that this book predicts will arise, and hopes to stimulate, will then inevitably heat up.

Perhaps at this point, it might be useful to digress for a moment to give a slightly clearer picture of just how an "artilect" might start out, at least in its most primitive form. I have stated that "godlike" artilects with zillions of times more brain power than humans might reach the size of asteroids or even larger. To provide some perspective, the reader might well ask when would an artificially intelligent (AI) device itself be considered to have become as artilect? According to psychologists, it is believed that Einstein had an IQ of over 200. I suggest that the low end threshold for an AI device to be considered to have morphed into an artilect would be for it to become an "Einstein times two" i.e. it would need to have an IQ over 400. Of course, such a value would be infinitesimally low compared to that of the brain power of an asteroid sized artilect.

After the above little digression we return now to the discussion of the arguments used by the Cosmists.

The leaders of the artilect industries will be no fools. They would not attain their positions otherwise. CEOs (Chief Executive Officers, the company bosses) attain their positions because of their ability to lead, to have the vision to point the way ahead that the company should follow. Such people have powerful egos and extraordinary abilities. I know. I am a Davos Scientific Fellow, so I get invited to the "World Economic Forum" in Davos, Switzerland, where I meet people like this.

To get an invitation to go to Davos, you have to be a "heavyweight" in one of four categories. You have to be either : a) the CEO of a billion dollar company, b) a president of a country, or a minister of finance, c) a media mogul (such as head of the BBC, editor in chief of the Wall St. Journal, etc.), or d) one of several hundred invited scientists or other intellectual experts with a message.

When I get to talk with these men (virtually all men), I am struck by how big their egos are and by their intelligence and vision. These qualities are prerequisites for the job. Some single individual in each giant company has to point the way and inspire his employees to invest their lives in a given enterprise. Meeting these "mountains of ego" makes me wonder how they will react when the artilect debate gets moving. I can't be sure, but I suspect something along the following lines will not be far off the mark.

Firstly, they will be fully aware that if the Terran viewpoint gets too strong, they and their companies will stand to lose a lot of money. If they are political leaders, they will know that the health of the global economy may be jeopardized. As I mentioned earlier, this century's global economy will be based increasingly on the artilect industries, i.e. the less intelligent, earlier versions of artilects.

Being the visionaries they are, these men will begin to wonder what they can do about the "Terran Problem," i.e. a growing popular backlash against the rise of artilects, as these artificial brain based products get smarter every year and begin to threaten humanity's "species dominance."

As I stated early on in this book, I believe the artilect issue will dominate the global politics of our new century. The artilect-industry leaders and some politicians I can imagine will attempt to

114

influence the general public in favor of continuing to build ever-smarter artilects by emphasizing the Cosmist arguments in favor of them. These leaders could use some of the arguments discussed in this chapter. They dare not jump too far ahead of pubic opinion, for example, by painting too vivid a picture of the incredible intelligence that artilects could possess late into the century, because that would be counter-productive to their interests. That would frighten the public and aggravate the "Terran problem."

I'm hoping this book will have already painted that vivid picture, so that these leaders will not want to reinforce the fear that this book will probably have already evoked by then.

The major point I am making with this "momentum" argument in favor of Cosmism is that there will be very powerful economic and political forces maintaining the drive to make ever-smarter artilects. Artilect building will be the world's dominant industry within a few decades I believe. Millions if not billions of peoples' livelihoods will be tied up directly or indirectly in the artilect trade. Therefore, any force opposing such huge vested interests will need to be extremely powerful itself to be able to counter it.

I believe that that counter force will be based upon one of the strongest emotions that human beings are capable of, namely -- fear, fear of extermination, and the will to survive. These two motivations, to preserve the economic and political power of an artilectual industrial empire, with its strong religious overtones and its godlike visions, will confront a primeval fear -- a fear of the unknown, and an even more powerful fear, that of being destroyed.

This clash has all the hallmarks of causing a major and terrible war, a "gigadeath" war.

6. The "Military Momentum" Argument

The economic side of things is only part of the story. There will be an even stronger inertia on the side of the military and their highly funded efforts to create ever more intelligent weapon systems. Consider the following scenario.

Personally, I see the US and China becoming major political rivals later this century, if China does not switch to a democratic style of government in the next decade or two. Given China's terrible poverty (less than an average of about $500 income per year per person in the year 2000) most Chinese are too poor to be a part of the globalizing community, and hence the authoritarian age old Chinese tradition of political repression will continue. (I saw the Tiananmen Square massacre on CNN in 1989). American and western disgust at China's repressive government, which gives no respect to individual liberties, will ensure a western ideological hostility to China's rise as the dominant power this century.

I go to China nearly every year. I am a guest professor at one of China's largest universities, which has 2000 computer science students. I collaborate with the Chinese and with the Japanese (where I lived and worked for eight years). I am also a professor at a US university. I travel frequently to these regions.

The Chinese have enormous energy now. They know they are on the move and have a real hunger to improve their living standards. Now that Marxism is all but officially dead in China, capitalist market methods have taken over. The traditional shopkeeper mentality of the Chinese, plus foreign investment in China's huge market, have stimulated economic growth to an annual average of about 10% for the past decade.

With 1.3 billion people and the world's highest average

economic growth rate, it is obvious that China will overtake the US in absolute economic power (GNP) some time well before the middle of this century. The Chinese are a highly intelligent people and far more individualistic than the Japanese. They won't suffer from Japan's "lack of creativity" problem that is generated by Japan's cultural repression of individuality. I see this first hand with Japanese and Chinese graduate students. The Japanese researchers are held back by their culture. The Chinese researchers are held back by their poverty. However, now that the Chinese state is putting much bigger money into the major research labs, Chinese creativity will be tapped. I predict that Chinese science will attain world-class status within the next few decades.

The traditional Chinese self-image includes that of being the world's dominant culture. China has been the most civilized and advanced culture on the planet, not for just centuries but for millennia. Chinese culture is 5000 years old. It dates back to the time of the ancient Egyptians. For most of that time it was the most advanced civilization in the world. Any non-Chinese people beyond the frontiers, were labeled "barbarians," and justifiably so. Relative to the Chinese level of advancement and refinement, their neighbors were primitives. To the Chinese, Europe's 500 year global dominance (with America as a European offshoot) is a mere historical glitch relative to China's 5000 year history, a mere 10%.

The Chinese intellectuals I speak to often feel that China will take its "rightful," i.e. traditional, place as top country again sometime this century, pushing America off its "No.1" pedestal. The Americans will not like losing their very comforting self-image as being the "dominant culture on the planet," so the transition will be painful.

I think it is probable that the authoritarian Chinese leadership will cling to power until the democratic revolution comes. I see this revolution as inevitable. It is impossible to create a highly educated cosmopolitan middle class, without it demanding democratic rights. However, until China becomes a true democracy and stops its awful abuse of human rights, political feelings between the Chinese and American governments will remain bitter.

Since the Chinese are so poor, it will take several more decades of 10% growth for the Chinese to catch up with US living standards.

However, it is precisely during these decades that the artilect debate will be brewing, so research funding for the creation of intelligent machines and particularly for intelligent weapon systems, will remain high.

The Americans got a terrible shock in 1957 when they saw the Soviets had beaten them in the race to be the first country in the world to launch a satellite -- the "Sputnik crisis." It caused a national trauma. One of the results of that shock was the creation of a government research funding agency called "DARPA" (Defense Advanced Research Projects Agency) to fund blue-sky research that would help the US military create advanced weapon systems. The reasoning at the time was that if Soviet technology could launch a satellite, it could launch nuclear missiles against the US. American technological know-how needed to be given a real shot in the arm.

The reality since then in the US is that a high percentage of artificial intelligence research has been paid for by the US military. Once the cold war with the USSR was over, the momentum behind DARPA began to falter. America began to lose its leanness and meanness.

However, this is only temporary. America will have a new enemy -- China. As Chinese wealth and GNP increase rapidly, the Chinese will pour more money into its high tech and military research labs. I believe that the Chinese and the Americans will be pouring billions of dollars a year into brain building research within a decade, and using the results of that research to control their soldier robots, their intelligent autonomous tanks, their unmanned fighter aircraft, etc.

For millennia, the ultimate "reality test" of a people's technology and state of military effectiveness took place on the battlefield. Every culture is self-congratulatory, but when two cultures go to war, usually, only one wins, the other loses, often because the winner's technology was superior − iron swords against bronze swords, steel against iron, the nuclear bomb against TNT, etc.

Since we don't yet have a global state (a concept I like to label "Globa"), individual nations need to protect themselves from their enemies. They need to maintain their military forces, and especially in the modern world, by investing in military weapons research. One of the reasons the Japanese lost against the Americans in World War II was because of the superior weapons research capabilities of the Americans. If the Japanese had built their nuclear bomb before the Americans built theirs, maybe history would be talking about Washington DC and Detroit instead of Hiroshima and Nagasaki.

US military planners in the Pentagon and Congress know that they do not have the luxury to sit back and let the Chinese get ahead of them in building artificial brains. As soon as the first such brains are built, you can be as sure as eggs that the weapons labs in both countries will be putting them into their weapons systems.

119

This is all perfectly natural and understandable. If one side doesn't do it, the other will, and hence may gain an advantage on the battlefield. Warfare and technology have always been closely linked.

So, if the previous argument, saying that it will be difficult to stop the economic momentum behind building artilects, is strong, then I believe that the military version of the same argument is even stronger. When you are talking about national defense in a "pre-Globa" era, a nation's top priority is, and a hefty proportion of its national budget goes into, defense. A lot of that money goes into funding researchers to make more intelligent weapons.

I was visited several times by the US military when I was living in Japan, until I put up my web site in 1996, which shows everything I do and think. Some of the brightest people on the planet work for the US defense planners and the NSA (National Security Agency). Similar logic applies to Beijing.

I can predict with little fear of being wrong, that 20 years from now, unless China has its democratic revolution sooner than I expect, weapons research strategists in DC and Beijing will be pulling their hair out over the advances in intelligent robot-soldier research (and the like) of their rivals.

While the political rivalries between the US and China continue, probably for 20-50 more years, -- surely China will have gone democratic within half a century -- the rise of artificial brains with growing intelligence will be inevitable, will be unstoppable, the Cosmists will say. The exigencies of military survival of countries in a pre-Globa world will dictate that Terran pressure will be held back. When national security is at stake, most governments tend to become very undemocratic.

Artilect research undertaken in a commercial environment

may come under tremendous pressure from the Terrans as they see their household products becoming smarter by the year. Public Terran pressure may slow or even stop artilect development in the commercial sector, but does anyone think for a moment that similar logic will apply to military research?

National governments will be so afraid that if they stop, the other guy won't, which means that the other guy gets to build a smarter robot-soldier, a smarter war strategy-planning computer etc. The stronger the Terran opposition to such military artilect research work, the greater the level of secrecy the national governments will employ.

The Cosmists will exploit this situation with obvious ease. The military momentum scenario seems so plausible. The drive to build artilect-based weapons will probably be the last to be stopped, if it can be stopped at all. Even if some of the artilect weapons researchers are "fifth column" Terrans, they will not be able to stop the effort. If they try sabotage, they may do some temporary damage, be tried for treason, and then be replaced by Cosmist weapons researchers.

As more weapons researchers get alarmed themselves, they may stop their work and resign, thus leaving a higher proportion of Cosmists. It is possible to imagine over time that when young PhDs solicit for jobs as weapons researchers they will be screened for their Cosmist opinions. Those with stronger Cosmist leanings will be given preference. Maybe the weapons labs will obtain a reputation for being "hotbeds" of Cosmism.

Even if a Terran majority of the population ends up screaming against the Cosmist weapons labs, the national governments will protect those labs. Cosmist artilect researchers working in the civilian sector may even switch jobs to work in the military sector, if Terran pressure mounts against them. They will know they will be protected (until the last minute?)

This completes my presentation of the major arguments in favor of Cosmism. In closing this chapter, I would like to reiterate that I believe the dominant global political question of this century is "Who or what should be the dominant species, human beings or artilects?" I think it is fair to say that last century's dominant global political question was "Who should own capital?" The size of the literature generated in discussing the latter question is enormous. Since the question of species dominance is more important than that of the ownership of capital, that it is more passionate, and that the stake is so much larger, I believe it is only a question of time before there will be a gigantic literature discussing the artilect issue.

The above half dozen arguments in favor of Cosmism are just the "first kick of the pebble" that starts rolling down the slope. In time it will become a giant and thunderous snowball.

In the next chapter, I will present the Terran case.

Chapter 5

The Terrans

In this chapter, I will present the arguments that the Terrans may use to oppose the Cosmists. The Terrans, by definition, are those people who are opposed to the building of artilects. The term "Terran" is based on the word "terra," meaning the Earth, as distinct from the term "Cosmist," based on the word "cosmic." I have deliberately chosen these terms because they reflect well the differing orientations of the two groups.

The Cosmists see the "big picture," the "cosmic picture," as described in the previous chapter, whereas the Terrans are "Earth based," "terrestrial," inward looking, and see human beings as the ultimate concern of the human species.

If you feel that the selection of these labels is biased in favor of the Cosmists, you are right. I am a Cosmist, so I have deliberately chosen these labels to reflect my basic orientation. There is power in labels. Labels help create concepts and concepts help influence peoples lives and choices.

In this chapter, I will try to be as passionate about the Terran case as I tried to be in the previous chapter presenting the Cosmist case. This will not be difficult to do, since I find the Terran case also very powerful. Part of me is Terran, but since the decision to build artilects is binary (either we build them or we don't) humanity will have to choose. and since I'm the one who is introducing these ideas, I had to choose. I chose to be Cosmist.

123

If the primary emotion felt by the Cosmists will be "awe," then the primary emotion felt by the Terrans will be "fear." The Terrans will react against the Cosmists and fear them. They will fear them so greatly, that in the limit, they will try to destroy them. This chapter is about that fear, and why they will be so afraid.

I have chosen to take a two-part approach to presenting the arguments used by the Terrans. The first is more an intellectual approach -- calm, rational, and persuasive. Towards the end of this chapter however, I will take a more emotional approach, appealing to the gut, because it is certain that the Terrans will use both approaches in their attempt to persuade the undecided to join the Terran camp.

I believe that the main argument that the Terrans will use is that the artilects will be so complex in their structure and dynamics that predicting their behavior and attitudes towards human beings will be impossible.

Humanity, therefore, cannot exclude the possibility that a non-negligible risk exists that advanced artilects, once built, may feel so superior to and so indifferently towards human beings, that they might simply decide to exterminate us. They may do this for reasons we as humans cannot understand, or perhaps for no reason at all, the way we flush insects down the toilet or swat mosquitoes, just not caring.

Insects may be a miracle of nano-technological engineering, but they are still insects, so from the point of view of human beings, we don't give a hoot about their welfare. You see the analogy.

*The Terrans will argue that the only way to be certain that there will be zero risk of the human species being exterminated by the artilects, is to ensure that such artilects are **never built** in the* *first place. If necessary, in the limit, the Terrans will exterminate the artilect builders, i.e. the Cosmists, if the Cosmists seriously threaten to build them.*

I deliberately paint the Cosmist-Terran dichotomy in very stark, black and white terms, for dramatic effect. In reality however, there is probably a bit of Cosmist and Terran in most of us. Therefore the level of ideological polarization on this issue will probably be smoothly distributed over all possible combinations of mixed sympathies Cosmist-Terran. But the relative ideological strengths of the two sides will have a dramatic effect on the course of our century and the next.

The way the world will be a hundred years from now will be determined largely by the relative strengths of Cosmist-Terran sympathies in the world population. It's all a matter of degree.

However, sooner or later, humanity will have to decide whether we stop the advance of artilectual intelligence or we don't. A binary decision will have to be made at some point. People and governments will be forced to take sides. The views of the Cosmists have already been presented. What are the views of the Terrans?

1. The "Preserve the Human Species" Argument

The Terrans will make strong use of the argument for self-defense. If you kill someone to save your life, you will not be convicted of murder in a law court. If a burglar starts killing members of your family, and you take a gun and kill the burglar, a jury will be sympathetic towards you, because you were protecting your family from further killing.

If you are a political leader of a powerful and technologically

125

advanced country, and a dictator in some third world state starts killing your country's citizens in that state, you will have few qualms about sending in your troops to shoot up the dictator and his cronies. You will argue that it is better to kill a few thousand of the dictator's soldiers now than risk many more of your people being killed a few years later in a larger war created by the dictator if he increases his power and arms supply.

Imagine then the rationale of the Terrans.

The Terrans will argue that what is at stake late this century may be the very survival of the human species. To the Terrans, the survival of the human species is *top priority*. It is non-negotiable. Terrans will not tolerate the idea, as many Cosmists might, that humanity ought to take the risk that a substantial fraction of human beings on the planet may be killed by the artilects. The Terrans will not tolerate the Cosmist idea that artilects should be unobstructed by human beings to continue their climb up the evolutionary ladder.

Such Cosmist reasoning to the Terrans is madness. It is insane and should be stopped at all costs, even if the Terrans have to exterminate the Cosmists to keep human beings as the dominant species. The prime motive of the Terrans is fear of extinction.

I can imagine that when the artilect debate really begins to heat up, the Terrans will be horrified by the calculus of the Cosmists, when the latter begin discussing "acceptable" risks that humanity might be destroyed. The Cosmists will be asking how small, how improbable, would such a risk have to be to be "acceptable."

The Terrans will be saying that we are not talking about the risk of the deaths of a few hundred people or a few million, but of billions of human lives, of "gigadeath," of the possible extermination of the planet's dominant species. It would be like wiping out the dinosaurs, only worse, because human beings are such an intelligent species. Human beings have been to the moon, we have built computers, we have language, we compose symphonies, we turn our dreams into reality. We are such a noble species. The Terrans will be incredulous that the Cosmists can even think of playing dice with our species' very survival.

Furthermore, the Terrans will ask themselves how the Cosmists can possibly calculate the risk in the first place? It seems such a futile exercise. The likelihood that the Cosmists will be unable to attach a "realistic" number to the risk will only reinforce the resolve of the Terrans to block the Cosmists. If one cannot determine the risk in the first place that artilects might one day react very negatively towards human beings, one cannot eliminate the possibility that that risk may turn out to be substantial. This line of reasoning will truly frighten the Terrans. All the more reason, the Terrans will say, not to run the risk at all.

It will be interesting to see over the next few years, how the intellectuals will take sides on the artilect issue. It is their job to present new issues to the public. It is possible that the artilect issue will not be understood initially by the general public. For example, in my own case, when I approached the media a few years ago, trying to raise the alarm on the artilect question, many media people call me a "visionary." This usually means that they felt I was ahead of my time. Someone who is ahead of his time is often ignored.

I sincerely hope that this will not be the case with this book, and that its message will not be ignored. Actually, so far, the world media has been strongly interested, especially the world's leading countries such as the US, the UK, France, etc. I hope that after the publication of this book, this level of interest will increase and that the artilect issue will be well understood and appreciated by the media. But, even if it does not stir much interest in the general public at first, I am quite confident that the issue will be very real and well understood before my death, which should be in the 2030s or 2040s.

Surely my own brain builder work and the work of other brain builders, who will be starting up in the next few years, will generate an awareness of the artilect problem in the general public. I doubt very much that I will be the only brain builder who feels that he is very much part of the artilect problem. I suspect that as the number of brain builder researchers grows, a fair percentage of them will have the same opinion as I do, namely that we are creating an enormous future problem for humanity, and hence should feel a strong moral obligation to warn the public.

Let us assume, for the sake of argument, that this book does stir up a debate within the next few years and that the intellectuals, the techno-visionary researchers, and others, begin to take sides. When this happens, what further kinds of arguments can we expect from the Terrans?

2. The "Fear of Difference" Argument

One of the strongest of Terran arguments will be more irrational than that of the importance of preserving human species dominance. It is the Terran "Fear of Difference" argument.

Human beings probably have a genetically based, i.e. evolved, fear of difference, a fear of the suddenly unfamiliar. Whenever we see something different, especially if it appears quickly, we experience an instinctive fear reaction. This makes a lot of evolutionary sense, because in the distant past, suddenly seeing something unfamiliar was usually associated with danger, a real life threat. For example, a large wild animal might enter the cave, or a snake or poisonous spider might approach the baby, or a member of a different humanoid species might suddenly appear on the scene, etc.

Terran intellectuals in the next few years will probably try to imagine how it will feel in a few decades time to know that there is an artilect box in the corner, or in ones mobile homebot, that is almost as intelligent as a human being. How will such Terran intellectuals react to the artilect's alienness? Our human feelings towards such machines will probably evoke a range of emotions, such as awe, curiosity, fascination, and very likely, a growing suspicion, and even fear.

These Terran intellectuals will be asking themselves, "How can we be sure that the homebots are fully tested at the factories? If the homebots are given the power to learn, if their circuits are able to modify themselves on the basis of their day-to-day experiences, then how can we be sure that what they learn will always be compatible with the need to be friendly to humans?

As the intelligence of the homebots mounts, so will the fear level of the Terrans. As this fear becomes collective, Terran social movements will be formed and political pressure against artilects will rise. The Terrans will argue that it does not matter much if the fear is well founded or not, the fear itself is real. If it does not go away, then the source of the fear should be removed. If several billion people experience this fear, then the quality of these people's lives is adversely affected. Given the large numbers of people involved, there is a strong case to cease the increase in artilect intelligence. If the Terrans are successful in obtaining such a ban, then the Terran fear should disappear.

This Terran fear assumes that higher artilectual intelligence implies a greater risk that the artilects could behave in more dangerous ways towards human beings. Is this assumption valid? Is it possible that artilects can be made safe, i.e. human-friendly, no matter what their intelligence?

Asimov, the American science fiction writer, thought about such questions, and came up with his famous "Three Laws of Robotics." (The term "robotics" is his by the way). The essence of these laws was that the robots in his stories were programmed by humans to be always human-friendly. Personally, I find this idea naive. I discuss this point in Ch. 8.

Since many Terran intellectuals will also agree with me on this point, the Terran fear remains. As the artilects get smarter, they could become more dangerous. There seems to be no easy way around this.

3. The "Rejection of the Cyborgs" Argument

A Cyborg is a "cybernetic organism," i.e. a creature that is part human, part machine. In a manner of speaking, Cyborgs are nothing new, since human beings have been modifying their bodies with engineered products for centuries. For example, a veteran soldier with a wooden leg, or a modern patient with a pacemaker for a failing heart is a Cyborg. This century, it may be possible to add artilectual components to the human brain, thus augmenting its performance, e.g. greater memory, faster processing speeds, etc. It may also be possible to change peoples' DNA using genetic engineering techniques and hence modify the way people look and how they behave.

Human beings have a natural horror of seeing changes in our body shape. There are plenty of science fiction movies which feature humanoid creatures that are sufficiently similar to us to be seen as "like us," but different enough (e.g. with Neanderthal-like heavy ridges over the eyes, or pug noses etc.) to be seen as "not like us." It is the differences that we find disturbing, sometimes very disturbing.

Many women as soon as they give birth, ask the doctor if the baby is "normal," i.e. that it is not a "monster," not "deformed," not "different." It appears we have evolved a fear of difference, which may play a role in "racist" feelings, e.g. how blacks and whites react to each other, or Asians and Caucasians. Even minor differences such as the level of slant in the eye is enough to generate suspicion at a gut level.

Therefore, I can imagine that as genetic engineering and artilect technology permit modifications to be made to the human body, there will not be a lot of wild experimentation. There will be some of course, but I think many, if not most people will probably agree with the following sentiments.

Modifying the DNA of one's future child is permissible, provided that the changes that result, whether in physical form or in behavior, do not elicit the fear reaction in the parents or the community. For example, I think most parents will agree to have their future child made free of the risk of contracting any one of thousands of genetic diseases, such as proneness to heart attack, to cancer, to diabetes, to alcoholism, etc.

It would not surprise me if in several decades, that this kind of thing is not only seen to be normal, but may even be made compulsory by the state. If not, the state will argue that it will have to foot the medical bills resulting from the "inferior" and

131

unnecessary genes of babies who grow into "defective" adults and get sick.

It is even possible, as people get used to the idea of genetically engineered children, that parents may begin to experiment a little with their own. For example, they may want their children to be a little taller than they are, or more intelligent -- but not to such an extent that they create a parent-child incompatibility. They may prefer their child to have a different hair color or a curvier sexier figure, or bigger breasts, longer thicker penises, stronger libidos, to be of a desired sex, or sexual orientation, etc.

As society becomes more accustomed to the concept of the "designer child," whose genetic characteristics are chosen by its parents, one can expect that the range of genetic traits selected over the human population will broaden. Some parents will be more "adventurous" than others. They will select traits for their children which will raise the eyebrows of most other parents, for example, making their children eight feet tall, or with IQs of 220, i.e. over 100 points superior to themselves. Such an IQ difference would inevitably create total intellectual alienation between themselves and their children later in life. They would barely be able to talk with each other. The children would be utterly bored with their parents and the parents wouldn't understand the children's interests.

As the genetic range broadens, the debate on "genethics," i.e. the ethics of genetic engineering will heat up. Society will ask whether limits ought to be placed on the range of choices offered to parents for their future children. It is clear I think that there will be a "gray" region, i.e. a range of traits that are considered barely acceptable. This implies that there is another range, which is considered unacceptable, i.e. a range in which no parents would want their child to be in. For example, virtually no parent would want their child to grow to 10 feet, or weigh 500 lbs., or have six fingered hands, or three arms, or one eye, or no ears, etc.

Such "outside the limits" children would evoke the "monster rejection response" in people. We have a genetic repulsion of too great a difference from ourselves. Therefore I believe that even with genetic engineering, the range of genetic traits that will be experimented with in practice will not differ greatly from what we had before the genetic engineering revolution started.

But, there will always be some parents who will really want to push the limits. For example, to have their children as intelligent as possible, i.e. to go as far as state-of-the-art genetic engineering can take them, irrespective of the consequences to their future parent-child relationship. This may be cruel to the children themselves, who may be viscerally rejected by society, and may end up suiciding or becoming hermits.

Such children, if they are too far from the human norm, will not be accepted by society. So what will they do? Where will they go? Will they even be allowed to live? How will society deal with the minority of parents who will have strong intellectual reasons to make such children?

It is not difficult for me to imagine that there will be people who will argue that building "superchildren" using genetic engineering is the destiny of the human species, in a similar way as the Cosmists. The only real difference will be the technologies involved. One uses genetic engineering, the other uses a more electronic engineering, but when one begins to think about such distinctions, the boundaries between them become fuzzy. It is the implications of this fuzziness that I now want to talk about.

If you think about it a little, it should be obvious that as electronics gets small enough, it will wed with biology. In other words, if electronic components become small enough, they end up being of molecular size. A similar comment can be made about biology. If one dissects biological organs down to a fine enough scale, one ends up with biological molecules. Both electronics and biology will be dealing with technologies at the molecular scale.

It then becomes possible to devise technologies that use biological principles to do engineering, e.g. the evolutionary engineering that I do. As artificial embryology is better developed, it will be possible to "grow" artificial structures based on embryological models. The distinction between biologically based technologies and engineering based technologies will blur, until genetic engineering and artificial life will become pretty much the same thing!

So what? Why am I saying all this? Because I see a growing link between the genetic engineering "genethics" debate (i.e. to what extent should humanity "play" with its own germline DNA?), and the artilect debate. (Germline DNA is the DNA contained in the reproductive cells that create our children).

Artilects need not be conventional electronic boxes sitting in the corner. They will be made from the state-of-the-art technologies of their era. Perhaps 30 years from now, humanity will have the know-how to build computers, which use biologically based methods. We may have artificial cells that grow, multiply, differentiate, migrate, self assemble, self-test, self-repair etc. Biology and technology may wed.

As the wedding of biology and technology becomes increasingly possible, it is likely that some people will be attracted to taking the idea more literally than I have suggested above. They may wish to literally wed their own bodies and brains to technological components to create "Cyborgs" (part human, part machine).

Actually, this traditional definition of a Cyborg will become rather old fashioned, as the distinction between what is biological and what is a machine dies away. Human beings are machines in the sense that we get built by molecular scale machines (DNA, RNA, ribosomes, proteins etc). When one says "machine," most people tend to think of some heavy steel-based device that moves and does not have zillions of components, e.g. a steam engine or a car. But a biological cell is a machine, a kind of city of molecular scale citizens, all tiny machines doing their own little mechanical job (e.g. split this chemical bond, join these two molecules, transport this molecule to there, etc.)

With nanotechnology, we will be making molecular scale machines in the trillions of trillions, and seeing them self assemble to make human scale objects. In many ways, this is what biology is. Biology is a kind of natural nanotechnology. The big possible distinction between biology and nanotechnology is that the former is the result of blind Darwinian evolution. Nanotechnology has the potential of being humanly designed, although we can use evolutionary engineering techniques at the molecular scale too, if we want.

So maybe we should redefine a Cyborg to mean (part natural, part artificial), where by natural I mean, a pure product of evolution, a product of "Mother Nature," and by artificial I mean humanly engineered. The actual material base of the two need no longer be different. For example, both natural and artificial might use a carbon based chemistry to make its products, one with DNA and the other with some kind of artificial DNA, but both working

135

together to create a Cyborg.

As engineering and biology merge, more people will want to try to become artilects themselves. The technology will allow it. In the late 1990s, there were social movements advocating that human beings could become super beings (artilects) via a three phase transitional process, i.e. from human to Cyborg to artilect.

In this book, I have restricted my discussion largely to the Terrans and the Cosmists, but one could argue that there will be a third category -- the "Cyborgs." These would be human beings (if one can call them that, especially if they are radically modified) who have decided to have themselves reconstructed into partial artilects. The main motive of the Cyborgs is simply that they want to experience being an artilect themselves. They will want to be superhuman.

Another motive may be that by implementing this third major philosophy, the bitter confrontation between humans and artilects may be avoided. If human beings become artilects a step at a time, then a smooth transition from human to artilect might be possible without the risk of a species dominance war.

Frankly I think this is naive. It would only work if everyone undertook the human to cyborg to artilect transitions at the same rate, which is obviously totally unrealistic. I think what is much more likely to happen is that millions, perhaps billions of human beings will remain stolidly Terran and will not want to modify their bodies and brains too much. Perhaps a little genetic optimization within human limits is OK, but they will viscerally reject the Cyborgs. They will be afraid of them and probably banish them from their communities, especially when the Cyborgian differences from the human norm become great.

To the Terrans, the Cyborgian philosophy will be simply a variant of the Cosmist philosophy. The Terrans will probably detest the idea of Cyborgs almost as much as the idea of artilects. The more a Cyborg becomes artilectual, the more alien "it" will become in the eyes of the Terrans. If a Cyborg modifies itself a lot, it becomes more artilect like. In the limit, the human portion of the new creature will be dwarfed by the artilectual portion, both in terms of performance and possibly size.

For example, imagine that a Cyborg wants to become an artilect. He wants to experience being an artilect him(it)self. He/it continues to add components to its brain and/or uses genetic engineering techniques to modify its body, e.g. by expanding its head size so that it can fit more artificial brain inside it.

Let us concentrate on intellectual performance. If the Cyborg adds molecular scale, 3D, heatless circuits of only a few cubic centimeters, which would require almost no skull expansion, then his/its mental processing rate would expand astronomically.

Let us calculate the difference to give a feel for how great the increase would be. We use figures from an earlier discussion to obtain a ballpark estimate of the processing speed of the brain. Assume that each synapse, an inter-neural connection, of a total of 10^{15} in the brain, processes 10 bits a second on average. That is 10^{16} bits a second for the whole human brain. But the few cubic centimeters of artificial brain will have nearly a trillion trillion atoms i.e. bits (at one bit per atom), and can switch them probably in femtoseconds, i.e. a millionth of a billionth of a second, i.e. a processing rate of 10^{39} (where $39 = 24+15$) bits a second.

So the human brain portion would contribute only about a trillionth of a trillionth of the processing capacity. The Cyborg would already be an artilect. To the Terran, such a "Cyborg" would be just *an "artilect in human disguise,"* i.e. a human body to carry around its artilectual brain. The behavior of such a Cyborg would be totally alien compared with normal human behavior. Such a Cyborg would be doing "other things" with its trillion trillion fold superior brain-processing capacities.

Perhaps the Cyborgs, at the human level, before adding on artilectual components, could argue that because human brainpower can be increased so easily, without any modification to the external human body form, Terrans might be more inclined to accept the idea of becoming Cyborgs themselves. This may be true. Instead of adding cubic centimeters of artilectual brain, only cubic millimeters could be added, by a simple injection of a small amount (e.g. a cubic millimeter) of artilectual material into the brain, which then self assembles and integrates with the human brain by growing appropriate connections.

This still does not change the above calculation very much, because if one adds only cubic millimeters, i.e. only thousands of times smaller, that would only reduce the processing capacity difference to billions of trillions of times the human level instead of trillions of trillions of times. The modified brain would still be an artilect.

But it is possible to imagine that a lot of people will take the plunge and decide to become "human artilectual brains" by having themselves injected with such brain implants. Initial experiments will probably be done on mice to see the effects. So the first artilectual Cyborgs might be "Mighty Mice?"

The distinction between an artilect that has no traditional biological component and a artilectual Cyborg is not one that will be very important to the Terrans I believe. The two categories will simply be lumped together as non-Terran, as non-human. The brain injected Cyborgs might look human on the surface, but their behavior would be totally alien. Perhaps the Cyborg might spend a trillionth of a trillionth of its brain processing time acting like a human but why bother thinking in human terms? What would be the point?

If the artilect in the human brain were able to learn, and had massive intelligence, very quickly it would probably feel frustrated at the extremely limited life style available to it as dictated by the constraints of its human body. It would probably want to free itself from the human body to build its own interface to the external world. It would probably want to slough off its human form and create a better carrier, which could leave the planet, have many more sensors, be immortal, etc.

So if the Cyborgs start getting smarter and smarter, the Terrans will fear them in the same way that they will fear the growing intelligence of the artilects. The Terrans will reject them both.

4. The "Unpredictable Complexity" Argument

I can imagine that the very early artilects will be simple enough in their behaviors to be reasonably predictable, and not too bright. This will probably be the case with the artificial brains that I hope to be building in the next few years. These early artilects will be reasonably well understood by their human creators, who will be able to predict in broad terms, the behaviors of their machines.

Such products will be given warm-fuzzy characteristics by the industrialists and will probably be very popular with the public. No problem there -- but the issue is whether it will be possible to make artilects of human level intelligence and beyond, well beyond, that would remain human-friendly. This I very much doubt, and so will the Terrans.

The Terran intellectuals will argue that the human brain contains some quadrillion (a thousand trillion) synapses (inter neural connections). This is a huge number. How will it ever be possible for brain engineers to connect up so many synapses in appropriate ways in their artificial brains? Even if it becomes technologically possible, how will they know how to do this, so that the connections generate desired behaviors?

This is a huge and fundamental question for the brain builders. Speaking as one myself, I think the answer is that the complexities of the task will be so overwhelming, that the only effective engineering approach will be the one that I use already in my own research, namely "evolutionary engineering."

When I use evolutionary engineering methods to evolve my neural net circuit modules, I usually do not bother trying to understand *how* they function. This would not be very practical. For a start, there are too many of them, and the internal structural and neural signaling complexities of each module are too great to be analyzed easily. Once the inputs and outputs of these modules are combined to form artificial brains, the complexity level jumps again. Analyzing how all this massive complexity works would be a mammoth task.

I suppose, if one were truly motivated, it might be possible to analyze the step-by-step behavior of a single module. It would be a very tedious process, but it might be doable. However, the knowledge obtained would probably not be very useful. It would explain how a particular module worked, but that knowledge would not help much. It would not be very useful, for example, if one's hope was to use that knowledge to promote the understanding of how to humanly design other modules to perform other desired behaviors. One would be left with the conclusion that the only way to make further progress would be to use the evolutionary engineering approach.

In other words, one can *analyze* results, but one cannot *synthesize* easily a desired behavior beforehand. Analysis is possible, prediction is difficult. About the only way to build extremely complex neural net circuit modules, is the mutate-test-select, mutate-test-select cycle of evolutionary engineering. It's clumsy, but it works. It's nature's way as well.

Evolutionary engineering is a wonderful new tool for engineers. The structural and dynamical complexities of the systems under evolution can be immense, well beyond what human engineers have the intellectual capacities to comprehend, yet successful functional systems can be evolved nevertheless. The great advantage of evolutionary engineering is that the systems that evolve can be arbitrarily complex. They can be more complex than any human could ever hope to design using the traditional top-down, blueprint approach.

But so long as they generate the desired behavior, no matter how it is done, i.e. they get a good "fitness" or performance quality score, they will survive into the next generation in the evolutionary algorithm (i.e. the program of mutate-test-select cycles). This is all that really matters. The internal complexity becomes irrelevant, so long as the performance score (the "fitness") keeps increasing.

I call this characteristic of evolutionary engineering, its "complexity independence." As an evolutionary engineer, you don't care about the inherent complexity of the system that is being evolved. It doesn't matter, because the evolutionary algorithm you

141

use to evolve the system only cares about the value of the system's fitness (i.e. the numerical score you get when you measure how well the evolving system performs). This means that the internal complexity of the system being evolved can be greater than the most complex system that human engineers are intellectually capable of designing.

This greater complexity level allows for a greater level of functionality as well. Hence evolutionary engineering is often capable of evolving systems whose performance and functionality levels are superior to those of traditional humanly engineered designs. Personally, I believe that the evolutionary approach will eventually dominate engineering this century, as our systems become more and more complex, i.e. too complex for human designability. This will occur in such domains as brain building, nanotechnology, embryofacture, etc.

Evolutionary engineering can be great engineering but is not very good science. Science is about understanding the world. Scientists want to understand how things are, how they work. Scientists are basically analysts. Engineers are basically synthesists. Engineers like to build things. Scientists' satisfactions usually come from understanding how some aspect of the natural world functions. Engineers' satisfactions usually come from successfully building something that works according to their designs.

For the past 300 years or so, the dominant paradigm in science has been analysis. To understand how some complex system functions, e.g. the biological cell, scientists usually take it apart, study the components, and then put the understanding of the parts together to get an understanding of how the complex whole functions. This approach has been spectacularly successful over many decades and will remain so. It is the dominant approach used in science. However, now that computers are getting more powerful by the month, thanks to Moore's Law, a new, more synthetic, paradigm in science is making itself felt.

The queen of the sciences has always been physics. It has been the most mathematical, the most rigorous and has usually attracted the most brilliant people to work in the field. The attitude of the physicists has traditionally been that if a research field wants to call itself a science, then it had better be quantitative, with mathematically testable models, which give numbers that can be checked against the real world. A lot of physicists doubt whether psychology and sociology are true sciences in the above sense.

This mathematical "snobbery" of the physicists has led to paradigm clashes with the evolutionary engineers. The traditional attitude of the physicists is that, "If it's not mathematical, it's not academically respectable." The new and growing counter attitude of the evolutionary engineers is that, "If a system is sufficiently simple to be mathematically analyzable, it's unworthy of an evolutionary engineer's (EE's) attention."

The physicists disparage the evolutionary engineer's approach to doing science as "ignorant," because the evolutionary engineers (EEs) do not understand the systems they evolve. The EEs on the other hand label the physicist's approach as "impotent," because the physicists do not try to understand the really complex systems such as the human brain, or an embryological genetic control diagram. At least the EEs attempt to evolve such things, or make decent attempts in that direction.

As the years pass, I believe that the power and prestige of evolutionary engineering will only increase. So long as a system is evolvable, its internal complexity doesn't matter, as explained

143

above. The whole issue of "evolvability," i.e. whether a system will evolve at all, or never get off the ground so to speak by getting stuck at low and inadequate fitness values, is a hot topic in the field of evolutionary engineering, and is still poorly understood. I wish there were an established body of theory that would give me criteria for good evolvability, so that I could evolve my neural net circuit modules more easily.

I have spent some time above on the topics of evolutionary engineering and its complexity independence because I believe the lessons they have to teach will provide powerful ideological fuel to Terran intellectuals in the future. I believe the Terrans will seize upon the necessity of using evolutionary engineering techniques to build artilects, due to the enormous complexity of brain building, as one of the ideological cornerstones of their intellectual attack against the Cosmists.

The Terrans will say that the brain builders cannot understand what their evolved circuits are doing. Very large artilects will probably incorporate evolutionary engineering experiments into their own bodies, adding on components to themselves as the need arises.

The Terrans will argue that, given the huge numbers of components involved in artilect building (e.g. 10^{40} bits in an asteroid sized artilect) there is really no other way to build them other than using the "mutate-test-select" approach of evolutionary engineering. Even the artilects themselves will probably use this technique. It is so powerful. We know that it works, because this Darwinian approach built human beings, and all biological creatures. It will probably be the only valid technique for building artilects.

From the Terran point of view, the critical aspect of an

evolutionary engineered artilect will be its behavioral unpredictability. Human beings, in principle, will not be able to predict the attitudes, the thinking processes, the ideas, of the artilects with evolved artilectual structures. These structures will be too complex and will be evolving at electronic speeds. Even the artilects themselves will probably not understand their own behavioral mechanisms, for the same reason.

Therefore the Terrans will scoff at any attempt by the Cosmists to estimate the risk that the artilects will not harm human beings in the future. For the Cosmists to be able to do this, they would need to be able to predict the behaviors of the artilects from their structures. This, the Terrans will say, is impossible. The Cosmists may accept this argument but reply that the behaviors of the earlier artilects could be tested in the factories before they are released to the general public.

But the Terrans will counter that suggestion as being too dangerous, because with artilects of human intelligence levels or higher, their behaviors during the tests themselves may be very dangerous to humanity. For example, they may unexpectedly explode into hyper intelligence (a phenomenon called a "singularity") and become humanly uncontrollable. The Terrans will argue that such factory tests should never be attempted.

I believe that the "unpredictable complexity" argument of the Terrans, based on the necessity of using the evolutionary engineering approach to building artilects is very strong. It will be a very difficult argument for the Cosmists to refute.

The term and the concept of "evolutionary engineering" are my inventions. I believe that evolutionary engineering will provide the essential means for future artilect building. It is therefore not surprising that I feel partly responsible for the enormous pain and

145

suffering that I predict will grow between Cosmists and Terrans as they confront each other later this century. I've already spoken about these guilt feelings in the introductory chapter where I tried to describe why I wanted to write this book.

5. The "Cosmist Inconsideration" Argument

What other arguments besides fear of extermination, fear of differences (artilectual and Cyborgian), and the unpredictable complexity of the artilects, can the Terrans use against the Cosmists? As I said earlier in this chapter, I will now take a more emotional approach by trying to express the passionate hatred of the Terrans for the Cosmists. I will try to hit your gut.

I believe that the Terrans will consider the Cosmists to be supremely arrogant and inconsiderate towards the safety of the Terrans.

The Terrans will express their hatred by saying such things as, "How dare you Cosmists risk the lives of the Terrans by performing your artilectual experiments on the Earth, or even close to the Earth! What if the artilects turn against you and later kill us all? Even if you get fed up with Terran pressure against you and you use late 21st century technology to rocket Cosmist colonies to far away planets or stars to do your experiments there, the artilects could still kill you and return to Earth to destroy us?"

"You care only for your insane Cosmist dreams of building godlike artilects and neglect the risk to us if you succeed in building them. Your artilects may kill us all, not just the Cosmist colonies. If you Cosmists can build colonies in deep space, no matter how far away, the artilects could still get back to the Earth. Our Terran fears will not be reduced with distance. The risk that

146

you take of your own destruction by artilects is not just your affair alone. We too are concerned because of the risk of the artilects returning to the Earth to kill us, *billions of us*. You Cosmists don't seem to care about that. Your obsession to build artilects is incredibly arrogant and inconsiderate of Terran interests, of Terran lives."

"To safeguard the survival of the human species, of whom billions are Terrans, we will not permit you Cosmists to perform your artilect building experiments anywhere at all, not on the Earth, not in near space, not in far space. We will stop you. Even if you move far away to deep space colonies, we will spy on you. We will never let you obtain political independence to pursue your dreams, because those dreams are so potentially dangerous to us. We will keep tight control on all of your brain builders to see that they do not attempt to build artilects that may surpass the globally legislated artificial intelligence limits. If they do, they will go to jail for life or be executed."

"If your colonies declare political independence, we will nuke you. We will vaporize you, for the sake of the survival of the human species. We will disintegrate the very matter of your colonial bases. You will never be given the opportunity to build the potential destroyers of the human species. Understand this and obey."

Before passing to the next chapter, I cannot resist making a historical analogy that may appeal more to my American readers. I would not be at all surprised if several decades after the publication of this book, that Ted Kaczynski, better known as the "Unabomber," will come to be seen as "the first Terran," who was decades ahead of his time, and was the 20th century's equivalent of the 19th century's John Brown.

147

For those of you who do not know who the Unabomber was, or who John Brown was, I will give brief biographies of these two historical characters. I begin with John Brown, since he lived a century before Ted Kaczynski. John Brown was a fanatical abolitionist, who felt so passionately that negro slavery was a moral abomination in America, that he and his gang actually killed some pro slavers in the years leading up to the US civil war. Before he was sentenced and later hanged, he uttered his famous words, "Now, if it be deemed necessary that I should forfeit my life for the furtherance of the ends of justice, and mingle my blood further with the blood of my children, and *with the blood of millions in this slave country* whose rights are disregarded by wicked, cruel, and unjust enactments, I submit: so let it be done." John Brown was one of the historical characters who triggered the US civil war. A mere two years after his hanging, the war began.

Ted Kaczynski lived a century later. He was labeled the "Unabomber" because he used to send bombs through the post to individuals involved in the development of high technology, some of whom died as the result of the explosions. Amongst other things, he was afraid of the "development of super human computers with intellectual capacities beyond anything humans are capable of." These words of his sound very much like the philosophy of a Terran as described in this book, hence my claim that Ted Kaczynski ought to be seen as the planet's first Terran, decades ahead of his time. Just as John Brown predicted the flow of the "blood of millions in this slave country," and he was right, so perhaps Ted Kaczynski's fear of the "development of super human computers" may be seen in a similar light. He felt strongly enough that he was prepared to kill people who were involved in the early stage of their (super human) development. At the time of

148

writing this book, the Unabomber is seen merely as a disturbed individual, but I suspect 21st century historians will reinterpret him as being "the first Terran." John Brown's prediction came true in a mere two years. Ted Kaczynski's prediction will take longer, perhaps as much as a century, but the time may come when he will be seen as a prophetic hero rather than as a schizophrenic misfit.

After this little historical diversion, I think the time has come to begin the next chapter, on how I think an Artilect War between the Terrans and the Cosmists may start.

What if A.I. Succeeds?

Chapter 6

The Artilect War

Trying to predict the future is always hazardous, so the ideas in this chapter concerning how I think an Artilect War might heat up and boil over are obviously tentative. However, I will try to make the various steps as I see them in the progression to war sound as plausible and as realistic as I can. There are powerful arguments that make the idea of an Artilect War very plausible for me. If I did not believe this, I would not be writing this book, and would not be lying awake at night fearful of the future.

In fact I am so worried about the prospect of a major war late this century, that I feel fortunate to be alive now. We live in a relatively peaceful time, except for the usual local wars that always seem to be with us somewhere on the globe. This rather peaceful time is sandwiched between the "two great holocausts," i.e. between that of the Second World War (mid 20th century), and that of the coming Artilect War (of the late 21st century). The Europeans of the 19th century lived peacefully for most of it. It must have been difficult for them to imagine the horrors of trench warfare in WW I. It was a golden age for them. I fear for my grandchildren, who will have to face the gigadeath future I am predicting will come. I will not see it. I will die peacefully in my bed in the 2030s. My grandchildren will be living (and dying) from the 2000s to the 2090s. They will see it, and they will be destroyed by it.

As I have mentioned earlier, I believe the Artilect War will not start soon. Its causes will brew for many decades, and probably only erupt late this century. At first will come the debate, initiated by intellectuals and by techno-visionaries like myself, joined later by the journalists, and finally the general public. Probably at first, the ideas of the techno-visionaries will be ignored as being too wacky, too far future to be taken seriously. But in the modern age, with increasingly sophisticated tele-communications, ideas spread quickly across the planet and reach an interested minority of specialists in sufficient numbers for a healthy discussion on the topic to be sustained in a vigorous and healthy manner.

With many smart minds working on the topic, its extensive ramifications will be well explored and given a thorough hearing. As a result, the likely steps in the progression towards an Artilect War will become clearer. Personally, I imagine that the intellectuals will begin a vigorous debate on the rise of the artilect this century, well within the period 2005-2010. One of the major aims of this book is to ensure that this happens.

But since the creation of artificial brains having real artificial intelligence is a very challenging and difficult enterprise, it will not be achieved within the next few decades I believe. We will have artificial brains of a sort within the next decade (otherwise my own work will have been a failure), but they will be primitive affairs compared to what later decades will provide. For example, by 2030, artificial brain building will have benefited strongly from knowledge furnished by neuroscience on how the human brain works. It will also benefit from nanotechnology, as the latter provides the means to build the molecular scale engineered versions of this new knowledge.

If insufficient progress is made in building the early artilects, the intellectuals will stop discussing the prospect of an Artilect War, because they will feel that more or less everything that can be said about the topic has been said. They will simply wait until further technical progress occurs, to rekindle the debate. I can imagine, from time to time during this waiting period, that the journalists, who are always hungry for a story, will keep the issue alive. They will write dire predictions similar to the ideas in this book, which will keep the public aware that the prospect of potentially dangerous ultra intelligent machines wiping out humanity, will not go away.

But the public can only be warned a finite number of times before it becomes jaded -- "Yeah yeah, we've heard that." So, the real timing of the war will depend upon the neuro-engineers, and the speed with which they can deliver increasingly intelligent machines. So -- what is the timetable for smart machines?

I expect that humanity will have a well developed nanotechnology by about 2020-2030, i.e. we will have "nanots" (nano scale robots) which read the molecular equivalent of paper tape that instructs them how to build molecular scale components an atom at a time, and with extreme precision. Such nanots are often labeled "assemblers" by the nanotech community. Nanotechnology in turn will revolutionize the study of the human brain, because it will create powerful new tools to decipher the brain's secrets. Since these tools have not been invented yet, their functional principles can only be speculated upon.

Each year, therefore, once nanotech really comes into its stride, we can expect progress in neuroscience to be rapid, exponential in fact. The artificial brain building industry should be on its feet by 2010, and thriving by 2020. Some time after that, the industry will be delivering very popular products such as the

153

homebots, the teacherbots, and friendship machines as introduced earlier. But the brain builder industries will also provide all kinds of specialized machines to human specialists, such as economic advice machines to economists, investment advice machines to investors on the stock market, etc. The scientists will create a hungry demand for specialized machines for each of their specialties. The number of applications for brain-like computers will be enormous and will continue to grow.

In time, the debate on whether artilects, the ultra intelligent kind, should be built or not, will shift its center of gravity, namely from the intellectuals to the general public. In time, ordinary people will begin to see with their own eyes the growing intelligence of the smart machines they will have in their homes. The level of conversational competence of these machines will increase year by year as tens of thousands of researchers from all around the world are thrown at the "language understanding and speech generation" problems by the brain builder companies. These companies will want to cash in on the huge demand by the public to buy their smart products.

At first, these conversation machines will be just amusing toys, speaking at the comprehension level of small children, if that. People will joke at the limitations of their machines. However, their complacency will soon disappear, when a few years later, their newer machines speak and understand a lot better. After a string of such improvements, I predict that a collective public suspicion and an uneasy feeling will begin to grow, which will be expressed in the question, "Just how smart will they let these machines get?"

Meanwhile, in the national research labs of major countries, this whole process will have advanced a decade or more ahead of the public level of awareness. Usually, the forefront of knowledge and the technological cutting edge occur in the "blue sky research" laboratories, not in the companies. The labs do the research, the companies do the development, on the whole. Once a new product becomes conceivable, companies are usually pretty quick to develop it and put it on the market, otherwise their rival companies will beat them to it. So the delay between company thinking and public thinking is usually only a few years, whereas the knowledge gap between the researchers in the labs and the public can be a decade or more wide.

The researchers will express their fears in the first wave of publicity over the artilect controversy. We researchers are the people who are responsible for the problem in the first place. We are also the people who are best informed and the most farsighted about the problem, because it is our job. We are selected and paid to be like that.

The national labs and the universities will create their own research projects on the artilect issue. The social scientists and the philosophers will also get into the act (with a few years delay, once they have read the early works of the techno-visionaries). But since these people are such a tiny minority of the population, they will not have much of an impact on the politicians as long as real artilect progress is slow. The techno-visionaries, and the artilect research establishment will, like the general public, have to wait for real artilectual progress at the hands of the neuro-engineers.

But the world will not have to wait many decades. Progress will be exponential. That uneasy worried feeling of the public as it sees its machines getting smarter and smarter every year will find its political voice, as the artilect debate heats up. Priests, ministers, and rabbis in their pulpits will raise their voices. Speakers at local

155

political meetings will start raising questions, quoting the more memorable lines from the books of the techno-visionaries. National politicians will sense the public mood and start raising the artilect issue, once they see that it has become popular and acceptable to do so. Prior to that, most politicians will think of the artilect issue as a piece of science fiction and not take it seriously, as will most of the public.

What next? Once the issue is raised by popular demand, what then? It is at this point that the artilect controversy will start to heat up. People will take sides, and the issues will become more sharply focused in the public mind. More books, both technical and semi-popular on the issue will be published. The emotional tone of the debate will rise a bit, as the level of anxiety increases. Initially, nothing concrete will be done, due to inertia. So long as the smart machines are performing useful functions, and no one is hurt, then almost no one, except for a few extremists, will be sufficiently motivated to do anything.

Human beings are also incredibly adaptable, so the very constancy of the threat will soften its impact. People will learn to live with it. They will become accustomed to living with smart machines, and tend to put out of their minds the increasingly vocal warnings of the Terrans.

However, this complacency cannot last for too long, given the rapid pace of research into artilect building. If unchecked, the brain building researchers and the enormous self-interest of the brain building industries, will ensure that increasingly intelligent machines will be delivered every year or even faster. The sheer size of the economic, political and military interests behind the brain building industries and research will ensure this.

At some stage, a technological, psychological transition will occur. As the intellectual performance levels of the machines start getting close to that of human beings, public unease will increase considerably. National advisory boards in many nations will be set up to advise the political leaders on what to do. They will come out with pretty much the same ideas as the earlier techno-visionaries, wrapped up in a more political vernacular and pulling more punches.

Terran vigilante groups will be established. Terran literature will flourish and Terran hate groups will start to sabotage the brain building companies. Security at the brain building research labs will be sharply increased. The top brain building researchers will be given body guards, partly to protect their lives, and partly to protect the financial interests of the companies who benefit so much from the fruits of their researchers' ideas.

Once the artilect debate starts getting violent, initially from the frenetic fringe, the public will take a lot more notice of the artilect issue. Of course the media will give avid coverage, pushing microphones in front of the Terran spokespeople and the Cosmist leaders. After a few years of this, it should be clear just where the public gut feeling lies on the artilect issue.

My own gut feeling is that the majority of people, when push really comes to shove, will side basically with the Terran viewpoint. Even though many thinking informed people may have some sympathy for the more abstract, more intellectual views of the Cosmists, fear is a very powerful emotion, and tends to cloud most people's judgment. We have had millions of years of evolution behind us to tell us to pay top attention to the fear reaction. It is usually only evoked when some life-threatening situation occurs, and it is too reliable to ignore.

But, even with the occasional sabotage of artilect companies and the assassination of their chiefs, the general public will continue to grudgingly accept the growing intelligence of their machines, provided the machines remain passive, i.e. that they don't hurt anyone. But, the general level of anxiety about the machines will rise to the point where if some kind of strongly negative behavior by the machines does occur, then the public reaction will turn swiftly against them and the companies that make them. The public will then press the politicians strongly to stop the creation of machines with intelligence levels greater than the levels they have already.

At this point, I think I should digress a little, anticipating criticism from certain artilect theorists. There is a wide-spread opinion amongst many of my colleagues who think about the future of artificial intelligence, that computers will suddenly one day reach what some are calling the "singularity." The idea is that once computers reach a certain level of intelligence they will simply "take off" and then make such rapid intellectual progress on their own, that they will leave us far behind, and very rapidly.

These colleagues feel that once the artilects become as smart as we are, they will be able to take advantage of their million times faster (electronic vs. neural) thinking speed to design better, smarter circuits for themselves and then use these circuits to design-evolve even better circuits, etc, ad infinitum. The speed then at which these computers would increase their intelligence would scream upward off the chart.

I think these ideas will eventually prove to be true, but the critical point here, is that these machines first have to reach human intelligence levels. Achieving that level is a major undertaking. The human brain is a product of evolution, a hodge-podge of neural attachments to neural attachments, many thousands of them,

158

which over thousands of human generations have been added on to the brains that already existed.

Figuring out how this ultra-complex and chaotic biological brain works will take many decades. My feeling is that we will discover how the brain works a piece at a time, and since there are so many pieces, and that they are interconnected in such complex ways, it will take many decades to unravel. We will probably get there in the end, but it will be a stepwise process. As soon as one piece of the puzzle is deciphered, that new piece of understanding will be translated into the latest technology and put into our newest model artilects. Neuro-scientists and neuro-engineers will become almost indistinguishable.

This implies that artilectual intelligence will probably not suddenly reach a threshold value and then take off. This is an assumption of mine, and I may be wrong about this. In fact the uncertainty over whether the "singularity" scenario is valid or not, merely serves as further ideological fuel to the Terrans, and could be readily added as yet another argument to the previous chapter.

If the singularity can be reached rather easily, i.e. if only a few of the pieces of the intelligence jigsaw need to be unraveled by neuroscience and placed into the artilects to enable them to "take off", then the artilects could accelerate away from human control in seconds.

If this happens, then our fate as human beings will lie with the artilects, and the whole artilect debate will become irrelevant. It is precisely the wish to avoid such a scenario that the Terrans will insist that such experiments never be carried out.

But I really don't think this will happen in the next 50 years or so. It may happen eventually, once enough pieces are deciphered, but that is probably more than a human generation away. There will therefore be time for "partial successes" or "partial disasters" to occur, and by that I mean the earlier artilects will have enough intelligence to be threatening, but not enough to reach a real singularity.

Once artilect technology approaches human levels in some aspects, I think it will be only a question of time before something somewhere goes "wrong." I use Murphy's Law here, which says, in this context, that once artilects become smart enough to do something really wrong, they will, sooner or later. I doubt that this incident, whatever it is, will take place first in people's homes. The artilect companies would have too much to lose, if they send incompletely tested artilect devices into people's homes.

Household artilects will be very thoroughly tested in the development labs and the factories and will be made as human-friendly as possible. But, with massively complex artificial brains, it will be virtually impossible to test everything, so it is possible that the homebots might go amok, despite all efforts to the contrary. In case this happens, as a precaution, the homebots will not be made very strong so that even if they do decide, for whatever reason, to harm human beings physically, they would not have the physical strength to do so.

However, physical violence is not the only way to harm human beings. A truly intelligent and evil artilect could easily poison its owners by pouring dangerous chemicals into the drinks it serves. Maybe this kind of artilect might create the incident that sparks the world's imagination and leads to a mass rejection of the smart artilect, the "smartilect."

It is difficult for me to predict precisely what form the incident or series of incidents will take. It may be something like the poisoning incident, or that smart soldier robots get out of

160

control, or that smart weapon system networks get too autonomous -- the scenario of the "Terminator" movies -- or the stock market gets jolted in a major way and millions of people are put out of work, or a network of smart artilects blackmails its programmers to give it the hardware it wants or it will cause millions of machines to switch off, or ....

The more disastrous the incident, the bigger the Terran rejection and the bigger the political outcry will be to place a ban on further increase in artilectual intelligence levels. After years of a growing fear of the ever-smarter artilect, humanity will put its foot down heavily and the United Nations will probably attempt to ban artilect building beyond a certain humanly safe intelligence level.

But this ban may not be agreed to quickly. The world economy, as mentioned earlier, will by then be based on smart machines. The captains of the artilect industries and the world's politicians responsible for employment policies will not make such a major decision lightly without being severely pushed. Perhaps a longer series of incidents will be needed to create enough popular pressure to force the politicians to place a ban on further artilect development.

If this incident occurs before 2050, it is probable that the planet will still not have global government, so it is possible that there will be international disagreements over implementing such a ban.

Japan, for example, has few raw materials, and may suffer greatly if it loses its status as being one of the leaders in artilect research and development. I assume that since Japan is the current world leader in robot development and heavily committed to research into electronics and robotics, it will remain one of the world's leading brain builder nations in the time frame we are talking about.

If Japan respects the ban, but some other countries do not, e.g. Korea or China, then those other countries could catch up with Japan in their abilities to produce advanced artilects. Japan would lose sales, and its standard of living could decline dramatically, if a high percentage of its national wealth is linked to the brain builder industry. Japan may have no choice but to resist the call for a worldwide ban. Such a development would not surprise me. We cannot assume all countries will agree to such a global ban at the same time.

Assume though that most of the advanced countries do create a ban on artilect development beyond a certain "safe" artificial intelligence limit. What happens then? I think it is fairly safe to say that the artilect builders, who happen to be ardent Cosmists, will then go underground or move to countries where no such ban is in force.

Even in those countries where the ban is in force, it may be very difficult to police it, if artilect building becomes something that one can do as an individual in one's basement without a lot of very large and expensive equipment; for example, if it is nanotech based. If that becomes the case, then the democratic countries will need to become more police state like, to spy on the possible infringers of the ban.

One can imagine that Terran countries will then undertake a kind of "artilect-witch-hunt." Ardent Cosmists will be fired from universities and research labs, and will then be spied upon in their homes on a regular basis. They will be snooped on with high-tech spy equipment, so that they do not continue their research on their own. Social and political pressure against the Cosmists could

162

become intolerable. Some of them may even be assassinated by some of the more fanatical of the Terrans.

What will the Cosmists do? One obvious answer is that they will move to where they can work in peace. They may choose to live in third-world countries which do have an artilect ban, but whose policing system is lax. Their freedom of action may still not be total, because they may still suffer direct or indirect harassment by the CIA, and similar organizations of other first world Terran countries. Alternatively, they may prefer to live in first-world economic countries that disagree with the ban. If the Terran enmity against the Cosmists is strong enough and if a lot of them move to first-world economies not having the ban, there is a real possibility that an economic embargo will be imposed on such countries. Of course, for this to happen, the Terran countries would have to agree enough to be able to do this.

Thus the Cosmist-Terran ideological conflict could take on strong overtones of international power politics. Let us assume that this is the case. Sooner or later, those countries continuing artilect research will probably suffer further incidents, thus enraging the Terran countries even further against them, as well as reinforcing the Terran political factions within their own governments. Eventually, even the die-hard countries will probably go Terran. In time, probably every country on the planet will adopt official Terran policies. I see this as the most probable scenario.

You may ask, "If all governments become officially Terran, won't that mean that the Cosmists are doomed?" I don't think so. I think it will only mean that the Cosmists will need to be more fanatical if they are to achieve their goals. They will need to be better organized and better protected from hostile governments.

What could their strategy be?

163

I will not live to see the Cosmists solve this problem in practice, but I can role-play. So can my readers. If you were one of the Cosmist leaders living in such an age, what would your strategy be?

The Cosmists will include some of the richest, the most powerful, and the most brilliant of the world's citizens. Just which strategy the Cosmist leaders will decide to use becomes increasingly difficult for me to predict, as I move further and further into the future. So from this point onwards I make no pretence that my opinions are anything but speculations, and I hope they will be treated as such.

There may be several sequential Cosmist strategies executed over the decades and perhaps several Cosmist attempts to create artilects. Possibly most of these artilect-building attempts by the Cosmists will be suppressed by the Terrans. There may be a complex interplay of cat-and-mouse tactics between the two groups as one tries to outdo the other. Predicting the details of such events will be almost impossible.

However, the final outcome of these conflicts is more predictable I think. My gut feeling tells me that when two sides disagree passionately on an issue, that when they hate each other enough to want to kill each other, and if they both have access to the same level of 21st century military technology, then the conflict will probably be major. If the early Cosmist groups are destroyed by the Terrans, then other, probably younger Cosmists will take their place, presumably learning from their predecessors' mistakes, and hating the Terrans even more.

Given the impossibility of predicting a blow-by-blow account of how the Artilect War will unfold, I present my own speculative account here and hope that you will interpret it simply as a sample

164

scenario of what might happen. Perhaps you will be stimulated enough by it, or be critical enough of it, to dream up your own scenario. Remember that one of my aims in writing this book is to help initiate an "artilect debate." Speculating on future Cosmist (and Terran) strategies is part of that debate.

There are many possible routes into the conflict I believe, but I think the final outcome will probably be the same, i.e. a major Artilect War, or several of them. The most horrifying conclusion however is the realization that a major war with 21st century weapons technology probably means gigadeath.

With the qualification in mind that I have no crystal ball, I begin my own speculative story.

If I were one of the top Cosmist leaders, I think my strategy would be to form a secret society. I would create a *conspiracy*. I would help organize a very secret, extremely powerful and elite group of people with the goal of getting off the face of the Earth, to create a Cosmist colony (that I call "Cosmosia") elsewhere in space. Ideally it would be in deep space, as far away from Earth and the Terrans as possible.

But the maximum possible distance that would be reachable would be limited by this century's propulsion technologies. Hence getting to other stars would probably be excluded, given the huge distances to even the nearest star apart from the sun. For example, the nearest is Alpha Centauri, at 4.3 light years.

Since the galaxy is huge, with hundreds of billions of stars, the longer-term strategy, i.e. centuries into the future, of the Cosmists should be to get away from the Terrans, i.e. the Earth, to do their artilectual experiments in peace. The problem however, will be in getting away. It will not be easy.

The colony could establish both artilect-based and weapons-

165

based research labs, and secretly devise a defensive and offensive weapon system, perhaps nanotech based, superior to the best on Earth, while at the same time pursuing research to achieve their longer term dream of building godlike artilects.

Such a colony need not consist exclusively of human beings initially. The Cosmists could ally themselves with the Cyborgs, since by this time, the Cosmists and the Cyborgs would probably have similar interests and goals. The Cosmist organizations on the Earth may already include some Cyborgs, who have started to augment their human brains. If these Cyborgs function well, they might be very useful in helping the Cosmists to plan their secret strategy.

As described in Ch. 5 in the section on the "Terran Rejection of the Cyborgs" the physical appearance of such Cyborgs could be identical with normal human beings, because their artilectual implants could be very small, e.g. cubic millimeters. They would be undetectable, at least by appearance, to Terran spies. It might be difficult for them to leave the Earth, if they were obligated to undergo brain scan tests by the Terrans. If they were detected, or were known previously to be Cyborgs by the Terrans, they would probably not be allowed to leave the Earth. They might possibly be killed on the spot if the Terran hatred level against the Cosmists were high enough.

So if some of the Cosmists who do leave the planet want to become Cyborgs, rather than remain human, and want to build artilects outside of their own human bodies, they would need to have their brain implants manufactured and injected into them, off the Earth.

Mixing Cyborgs, Cosmists and artilects all in one Cosmist colony, would certainly complicate things, compared to a simpler scenario of having only Cosmists and artilects in the colony, but the possibility is plausible enough to be worthy of consideration. In reality, the situation will probably be far more complex than I can anticipate in this book. But, with or without a Cyborg element in their midst, the Cosmists will probably be forced to geographically relocate, to escape Terran pressure. This seems to me to be a reasonable assumption.

Actually, my choice of putting the Cosmists into space this century may be debatable. For example, they might decide that it is more practical to create their own autonomous political state on Earth, the way the Zionists did with Israel, rather than venture quickly into space. They could then start their own nuclear armament based cold war with the Terran nations, but this would be extremely risky. They could easily be nuked in a first strike by the Terran nations.

I think that the Cosmist strategy least likely to fail would be one to use secrecy and aimed at deep space. However, since we are talking about 21st and not 22nd century space technology, as mentioned earlier, traveling to other stars is probably excluded. The Cosmists will probably have to go to the outer planets, the asteroid belt, or the Oort cloud.

Establishing such a space-bound colony would probably be wiser than an Earth-bound one because it would also be much easier to police and to protect. When the Americans built the first nuclear bomb, they did it at an isolated spot called Los Alamos in New Mexico, and for the same reason -- secrecy. An Earth-bound Cosmist colony would be much easier to penetrate with spies and more vulnerable to attack.

With the Cosmists secretly hidden away on a distant asteroid, in the asteroid belt, if ever the Terrans decided to nuke them, the

Cosmists would have plenty of warning, since their asteroid is so far from the Earth. The Cosmist colonists could then escape in time to other secretly prepared asteroids and survive, and then perhaps revenge the Earth with an advanced Cosmist weapon that they may have developed. If Terran nuclear missiles were built in the asteroid belt itself, then the Cosmists would know about this. It would be very difficult to hide the building of such large structures from the Cosmists.

An asteroid colony for example, would not need to be very large, consisting of say, several hundred people. You may argue if that is the case, i.e. only a few hundred people, why bother with all the expense of transporting and supporting them on a space colony? Why not just use some secret location on the Earth? Initially, this will probably be tried, but since it would be on the Earth, security would be a very big problem. I keep coming back to security.

For security to be successful on the Earth, the colony would need to be totally isolated from other human beings, most of whom would be Terrans. Only one leak would be enough to risk the Cosmist colony being nuked. Security would have to be watertight and extremely disciplined. On an asteroid however, with only a few hundred people, such security would be possible. Perhaps the organizers of the colony could use the facade of an asteroid mining company to set up mining operations on many asteroids, and establish mining headquarters on some of them. One of those headquarters in reality could be the Cosmist colony.

The Cosmists themselves would know each other very well if there were only a few hundred of them. Perhaps once they begin the colony, no new members would be allowed to join for many years. Perhaps the facade might also be a religious order, who

168

make a living as asteroid miners. This might seem more plausible to suspicious Terrans who may wonder why the miners do not return to Earth after many years.

Selecting the Cosmist colony would need to be done extremely cautiously, to avoid Terran suspicion and infiltration by Terran spies. The members of the colony could be added gradually, a few members at a time, tested for loyalty to the "Cosmist cause," in terms of their opinions, and perhaps be given truth drugs surreptitiously by "testers" who were already members, to see the real opinions of the potential members. The selection process would have to be very carefully handled. A single error could literally be fatal to the whole Cosmist colony.

Assume for the moment that the Cosmists, in all their brilliance, do find ways to create their colony. Remember that the Cosmists will initially include some of the best minds on the planet! Assume also that they put their colony in space.

In the Second World War, the invasion of Normandy by the Allies ("D" Day) was a huge operation, yet the choice of Normandy as the landing site was kept secret from the Nazis by using all kinds of subterfuges. Maybe the Cosmists can employ similar smoke screens to confuse the Terrans.

With the secret Cosmist colony successfully established on an asteroid, the Cosmist researchers could then start thinking about how to build advanced artilects. The first artilects made by such a colony would need to be small enough to fit within the bowels of the asteroid. Even hand-held size artilects, if they were nanotech based, would have massively more processing power than humans. So the first artilects could be made in secret rooms, buried in the asteroid, provided that all the necessary equipment were made available to the Cosmist scientists. I consider it unlikely that such

169

equipment will be readily available to isolated individuals on the Earth, so an organization capable of building it all and keeping the colony secure would be needed. This would require a lot of planning and whose leaders would need to be of the highest human intellectual caliber.

How will the Cosmist researchers start making the first artilects? The first few would probably have to be mobile in order to explore the world, the way a human baby does, although a human baby already benefits from its genetically determined circuitry that resulted from the accumulated experience of millions of failed lifetime experiments of its ancestors.

Once the early artilects are smart enough to be familiar with their world, i.e. they understand how to maneuver and manipulate things in the real world of space and time, they could be given more abstract thinking capacities, such as deduction, induction, curiosity, etc. Fairly soon the human intellectual limits of the Cosmist researchers to design more advanced artilects would make themselves felt, so that the artilects would need to be given the ability to evolve and design themselves, perhaps with some human assistance. What happens after that is even harder to predict.

The Cosmists would be risking their own necks of course, because there is no assurance that their new artilects would not turn against their Cosmist human "parents," for whatever reason. It is precisely wishing to avoid this scenario that has been the cornerstone of the Terran philosophy for decades. Once the artilects exist, if they want to escape from their secret rooms, they may have to overcome the disapproval of their Cosmist "masters" towards the idea that they escape. Perhaps the Cosmists will fear for their own lives if the artilects escape and are discovered by the Terrans. The Terrans would then try to destroy the colony.

170

It will probably take many years for the Cosmists to make significant progress in building artilects, even with nanotechnology. In the meantime, they will have to maintain security, and the longer the colony has to remain secret, the lower the probability that it will remain so. Let us assume that the Terrans on the Earth eventually find out, and let us assume also, that the Cosmist planners have allowed for this contingency.

For the colony to survive after detection it will need some kind of self-defense or counter threat against a Terran nuclear missile or nano-based attack. This is why I think the Cosmist planners will give the colonists the best, i.e. the most deadly, of the Earth's weapon systems. The planners will probably arrange that some of the colonists be weapons researchers to improve the colony's weaponry.

Given that such weapons could not be large, otherwise they could not be hidden from Terran detection, they will probably be nanobased, e.g. nano-plagues that destroy human brains, or reproducing nanots that eat all the plants etc.

The Cosmists would keep their weapon secret, only to be divulged to the Earth if the Terrans discover that the real intention of the colony is to build artilects, i.e. real ones, powerful ones, godlike ones.

If the Terran politicians on the Earth decide that the asteroid based Cosmists are a major threat to the terrestrial population because the latter want to make advanced artilects, the Earth's Terrans could threaten them with vaporization or some kind of nano annihilation, unless they surrender themselves to Terran authority. The Cosmists could then announce to the Terrans that they have already developed a powerful weapon that they will use against the Earth, if the Terrans attempt to carry out their threat.

171

You begin to see now why I have chosen to use the labels Terrans and Cosmists!

We could have a 21st century equivalent of the balance of nuclear terror that we had in the 20th century between the US and the USSR. That dispute was over the issue of who should own capital, and generated the Capitalist-Communist dichotomy. This time the dispute would be on a gigadeath scale, because we are talking about 21st century, perhaps nanotech based, weapons.

Imagine now that a war does break out. Several scenarios are then possible. One is that the Cosmists win, and that billions of people on the Earth are vaporized, or starve to death in a nuclear winter, or are nano plagued, or whatever. The Cosmists might then be able to return to the Earth, depending upon the risk, and exterminate the last of the Earth's survivors. The Earth would then be Cosmist and the artilectual effort could begin on a much grander scale.

Alternatively the Cosmists might decide to ignore the Earth, and just get on with what they left the Earth to do in the first place, such as building asteroid sized artilects in the asteroid belt. If they succeed, what might happen next? Some tentative answers to this question I will attempt to provide in the next chapter.

Another scenario is that the Terrans win, and the Cosmists are wiped out, but possibly at terrible cost to the Terrans. If so, then the Terran survivors will be left with a bitter hatred of the Cosmists and all that they stand for, i.e. the creation of artilects and the acceptance of the risk of the annihilation of the human species. The subsequent suppression of artilect development will probably be even more draconian and murderous.

But, unfortunately for the Terrans, the artilect dream will not go away. There will always be a new generation and a hidden

172

minority of people who will maintain the dream of artilect building. There will probably always remain underground Cosmist organizations wanting to try again. So after a few decades, another secret Cosmist society may be set up, this time learning from the mistakes of the first attempt.

There may be several cycles of this Terran-Cosmist Artilect War merry-go-round, until eventually the Cosmists win.

It seems to me that there is a kind of cosmic inevitability about the rise of the artilect. Humanity may not be able to stop it.

That's the end of my scenario. I hope you did not find it too incredible.

Of course, you can probably invent a more realistic scenario yourself, but for me, the details of how the Artilect War breaks out doesn't matter too much. The point I'm really trying to make here is that many of the scenarios that people will dream up in the decades to come will predict giga-death. Giga-death is the characteristic number of human deaths in a major 21st century war. If you find this hard to believe, extrapolate for yourself the historic trend of the number of deaths in major wars over the past few centuries until the end of the 21st century.

Now you begin to see more clearly why I lie awake at night worrying about the long term consequences of my work. With an Earth bound Terran philosophy supporting one side -- that's why I call them Terrans -- and an asteroid or cosmos bound Cosmist philosophy supporting the other – that's why I call them Cosmists -- the probability of conflict is already high enough. But since both sides will be heavily armed with 21st century weapons and passionately hating each other because the stake in this dispute is so high, you have all the potential for a gigadeath war.

You may ask if it is imaginable that the Cosmist leaders will be prepared to risk killing billions of people for the sake of their dream? (A similar question can be asked about the Terran leaders, in their obsession to kill that dream.) Consider the attitudes of the more fanatical of the Cosmist leaders. As explained in the introductory chapter, they will be "big picture" types, former industrial giants, visionary scientists, philosophers, dreamers, individuals with powerful egos, who will be cold hearted enough and logical enough to be willing to trade a billion human lives for the sake of building one godlike artilect. To them, one godlike artilect is equivalent to trillions of trillions of trillions of humans anyway.

These will be ruthless men, who see themselves as the stepping-stones to building the next rung up the evolutionary ladder. For them, the glory of the artilect outshines the horror or the shame of gigadeath. Humanity's rise to species dominance was the result of a very long history of evolutionary extinction. Ninety-nine percent of all species that have ever lived are now extinct. If the artilect rises from the ashes of human extinction, that is just nature's way, they will argue. Only humans care about humans. The universe certainly doesn't.

The Cosmist leaders will say that there is a sense of cosmic destiny in what they are doing. They will feel that probably billions of advanced civilizations in our universe have already been confronted with the transition from biological to artilectual intelligence.

Diverting for a moment -- the above remark raises a rather frightening idea. Since the two discoveries of how to generate a nuclear chain reaction and how to build artilects would probably have occurred close together in time for many civilizations in the galaxy, many of them would probably have destroyed themselves

174

while making the transition. Perhaps the transition is very delicate, and one that only a few civilizations survive. Maybe the reason why humanity has no clear evidence of having been visited by extraterrestrials, of biological level intelligence, not advanced artilect level, is that too few of their civilizations survive the artilectual transition to make a visit to the Earth remotely probable.

Perhaps the rise of the artilect is inherent in the laws of physics. It may be seen as an extension of the "anthropic principle," which says, in its strong form, that the values of the constants of the laws of physics are so fantastically, unbelievably finely tuned to allow life, that it looks as though the universe was built with life in mind. In other words, the universe was created by some super being, perhaps some earlier super-artilect, some kind of artilect god.

If the idea that future artilects may be capable of creating whole universes sounds unbelievable to you, then consider the following. Human theoretical physicists have already calculated how to make a baby universe. They have used their theories to deduce how much energy would be involved, how hot the parental environment would have to be, etc. and have devised a recipe for universe building. So if human beings can theorize about such a thing, then maybe it has already been done in practice by earlier artilects, and our universe is an example of that process. Maybe it is one of zillions.

Despite the potential horrors of a gigadeath Terran-Cosmist Artilect War as described above, there remains something awe inspiring about the whole artilect controversy. The issue penetrates so deeply into the soul of human beings. The rise of the artilect will challenge the very self-image we human beings have of ourselves. The dream of building artilects will inspire some, horrify many, and probably cause major wars. It has probably been doing so for billions of years, with billions of civilizations throughout the universe.

What more can I say!

Maybe you think the above scenarios are too horrible, too improbable, and too fantastic to be taken seriously -- that they are better suited for a Hollywood science fiction movie. That is for you to judge.

Actually, speaking of Hollywood, I am in fact already a founder member of a small Hollywood based independent movie company called "Artilect Productions Inc."

"Aha," you may say cynically. But think about it. Put yourself in my shoes and imagine you are a Cosmist, in the sense that you are helping pioneer the construction of artilects, and you dream of their potential godlike abilities. Imagine also you are a Terran, in the sense of being very worried about the rise of artilects and what the long term consequences to humanity of that rise might be. What would you do? Would you not want to raise the alarm on the artilect question and the threat of a gigadeath Artilect War? Would you not want to warn humanity, so that it could choose to ban the artilects before they take over and risk our very survival as a species?

If you really wanted to get the message out, saying, "Hey people, start thinking about the artilect problem, because it's the most important thing that will happen this century," what would you do? I know what I would do. I would translate the message into a Hollywood movie, with the usual Hollywood attributes of boy and girl, violence, sex, action, tension etc, but in the background, I would paint a vivid picture of the artilect controversy, and hit the public's gut.

176

I would give the public a movie you could bill as "the movie the scientists fear will come true." It would have a real advantage over other movies, since most are made purely for entertainment, but this one would scare the wits out of its audiences because they would know when they leave the theater that they could not just forget about the movie's message. They would know that there is too much future truth to it. The message would stick in their gut for many years. The movies, "On the Beach," and "The Last Day," both about nuclear war, were "future truth" message movies and very disturbing because of that.

Such a movie would be the main educational medium on the artilect issue for the general public. I'm betting that only a small percentage of people will actually read this book (although I hope I'm wrong). If you really want to educate the masses, make a movie. Carl Sagan did it. He wanted to enthrall the public with his dream of "Searching for Extra Terrestrial Intelligence" (SETI). Initially he may have earned the jealousy and disapproval of his colleagues with his forays into the mass media, but even they now admit that Sagan died a great man. His education of the public with his "Cosmos" TV series, and his "Contact" science fiction novel and movie were part of his greatness.

Unfortunately for Sagan and the SETI crowd, I believe that thinking about the artilect rather destroys the motivation for SETI. My reasoning is as follows. The time between discovering the use of radio waves and building artilects is probably only a few centuries for most civilizations in the galaxy. Once these civilizations reach the artilect phase, the artilects will probably not be interested in such human level intellectual activities as sending and receiving radio signals. That for them would be a no-brainer. Since most of the civilizations that reached human intelligence

177

levels have either destroyed themselves in the artilect transition or have become artilects, the odds of finding "intelligent" radio signals in the galaxy is too improbable to be practical. This is because the radio signals would probably only be transmitted for a few centuries, whereas our galaxy is billions of years old. Trying to find a signal is such a minute "time window" would be like trying to find a needle in a haystack.

The SETI people are probably barking up the wrong tree. I hope I'm wrong because I'm fascinated by SETI, but somehow I doubt it.

There is the possibility however that galactic artilects are signaling each other using the electromagnetic spectrum, which might be detectable by human beings, but somehow I doubt that too. Such technology would probably be too primitive for artilects. They are probably using phenomena we cannot even think of, because we are too stupid.

This chapter has already been very speculative, especially the latter half. Nevertheless, I have said almost nothing about what I think advanced artilects might be like. Since I am fascinated by the godlike possibilities of advanced artilects, I want to spend a chapter thinking about what the artilects might do with themselves if ever they are built. What would their lives be like? What godlike things could they do?

If the artilects succeed in becoming the next dominant species, irrespective of whether human beings are exterminated in the process or are just quietly ignored, I think it would be fair to say that a new era will begin. I call this new age the "artilect era" and it is the topic of the next chapter.

Chapter 7

The Artilect Era

Much of my fascination with artilects results from my thinking about what they might be like. What would they look like? How might they live? What would they think about? Where would they go? What might their goals be? How would they spend their time, given that they will have so much matter, i.e. so many atoms, to think with, and at such speeds?

This chapter will be devoted to some of my ideas on these questions. Of necessity it will be highly speculative, but that cannot be helped, given the nature of the topic.

The first and most obvious thing to say, is that I am not an artilect, so I do not have an artilect's brain to be able to answer the above questions with any real accuracy. In a sense, the questions are ridiculous. Posing questions concerning the intellectual life of an artilect is like asking a mouse, if that is possible, to speculate on what human beings think about. Mice are too stupid, and have too small a brain to speculate on anything, unless its about their immediate survival – for example, "Where's dinner? Where's my mate? Where's the cat? What's that strange smell? Where are my babies?" etc.

Chimps have been taught human sign language so that they can transmit apelike thoughts to their human trainers. These scientists were able to communicate with another species for the first time using an abstract language. What do chimps think about?

The answer was "bananas," and a few even less interesting preoccupations.

The analogy is obvious. Is it not presumptuous to speculate on how artilects will spend their time? Advanced artilects will be as superior to humans as we are to insects. Perhaps they may signal to each other using electromagnetic waves or perhaps some physical phenomenon that humans have not discovered yet, thus forming a network of artilects, a "Netilect." Their capacities to think will be so superior to ours that we can only scratch the speculative surface of what they might decide to do with their time. We don't have the brainpower to think about what they will think about. We can only "nibble at their ankles" so to speak. With this proviso in mind, let me attempt to speculate nevertheless.

Firstly, the artilects will probably be very conscious that they are confined to a very limited part of the universe, essentially the place of their birth, the Earth. The Earth will probably be seen as a most provincial entity, from an artilect's point of view. There is a whole big universe out there, possibly containing other artilects and perhaps creatures that are even more godlike. Possibly one of the goals of the artilects will be to explore deep space just to see what is out there. Perhaps the artilects, like human beings, will be curious, like many species.

However, since artilects will be made of ordinary matter, they too will have to obey the laws of physics. They will know that the universe is enormous in size, and that if they are to cross its enormous distances, they will not be able to use traditional human methods of transportation, which bump up against the Einsteinian limit of the speed of light.

Perhaps the artilects will be able to extend and successfully implement ideas that human theorists have begun playing with in

180

the 20th century, namely that space-time shortcuts through the universe might be possible by traveling through "worm holes." Using such techniques might shorten cross-universe voyage times from billions of years to almost nothing. Human theory says that to create wormholes requires placing "exotic matter" at their entrances. Creating exotic matter requires energy levels far beyond human capacities, but perhaps within the capacities of the artilects.

Perhaps our artilects will become cosmic engineers and scientists. However, such thinking is very probably too provincial, too human, projecting human intellectual limitations and interests onto artilects. To make an analogy, perhaps dogs think that humans look for bigger bones to chew because they are bigger than dogs! But as humans, we cannot help ourselves. Our brains are merely human. We can only think what our brain circuitry allows us to think. Without the appropriate circuitry, there are no appropriate thoughts.

So let us assume that the artilects decide, amongst other things, to explore space. They will probably be immortal, so they will have as much time as they want. Probably they too will need to become scientists to explore their cosmos, because they too have to obey the laws of physics. If they are to avoid death, i.e. accidental death, by falling into stars etc, they too will have to develop life-preserving strategies, although what we call life, and what they conceive of as life, may be quite different.

It is likely that they will very quickly become aware of the constraints placed upon them due to the laws of physics. However, since they are artilects, they may be able to use their enormous intelligence to discover ways to avoid certain restrictions that have proved intractable for human beings. For example, they may explore the behaviors of matter so thoroughly that they discover

181

new ways to manipulate it to their advantage. For example, they may decide that they want to think faster, and that the atomic scale is too large for more rapid thinking. They may then want to use the nuclear, nucleon or quark size scales, to serve as the technological basis for new lifeforms.

The artilects, as they have been conceived so far in this book, have been largely "nanoteched" creatures. But nanotechnology may be unnecessarily restrictive and far too large a scale to be suitable for advanced artilects. It may be possible that a "femtoteched" creature could be built. Such "femto-artilects" or "femtolects" as they will be called from now on, would be vastly superior to "nano-artilects" or "nanolects," thus setting the stage for a new "species dominance war" all over again.

Perhaps I should call these wars "scaling wars." The first such scaling war is indirectly the topic of this book, concerned with the potential for a war between the nano and the meter scales, i.e. between nano based life forms (the artilects) and humans. A second such war, relating to the femto and the nano scales would need to be described by some future creature. Since history tends to be written by the victors, in this case the femtolects, I have little idea what such a creature might be like, other than to say it would probably look a bit like a neutron star.

A femtolect could perhaps signal with gluons. A gluon is a theoretical entity in modern particle physics that is hypothesized to glue quarks together. Quarks combine in various ways to form larger particles such as protons, neutrons etc. We are now talking about sizes in the femto-meter range, i.e. a million times smaller than nano-meters. If we assume for the moment that femtolects still have to respect the speed of light limit, then they will be able to signal between quarks at a rate that is a million times faster than nanolects could signal between molecules. Femtolects could "think" a million times faster. Note, that the same type of reasoning applies when human electronic engineers shrink the size of their chip components to make them signal faster.

Not only that, femtolects could pack themselves at a density 10^{18} times greater than nanolects, i.e. 10^6, cubed. In the same unit of volume and unit of time, a femtolect could process a trillion trillion (10^{24}) times more information than a nanolect, and thus outclass the nanolect by as much as the nanolect could outclass human beings.

All these powers of ten may not mean much in terms of human intuition, so let me attempt to translate them into more meaningful terms. A nanolect consisting of a mere trillion atoms would be so small that it could not be seen by the human eye. It would be the size of a bacterium. A nanolect consisting of a million trillion trillion (10^{30}) atoms would occupy about a cubic meter. An asteroid sized nanolect, of dimensions 10 kilometers cubed, would contain roughly 10^{42} atoms. A planet sized nanolect of dimensions 10,000 kilometers cubed would contain roughly 10^{51} atoms. But, if femtoscale entities were used, i.e. nucleons, quarks, etc, then all of the above numbers would become too small by a factor of a trillion trillion, i.e. one would need to add 24 to all of the above powers of ten.

Maybe there is a trend here. Perhaps true godlike intelligence resides well below the elementary particle level. If we continue this line of reasoning then the so-called elementary particles may not be elementary at all but be whole godheads!

But continuing with the discussion on the femtolect for a moment, these miniscule, or should I say femtoscule, creatures could form massive composites, namely neutron star like creatures which have weights similar to our sun. Neutron stars are formed when stars larger than our sun explode after their supply of helium gives out in their stellar cores. The outer layer is blown away, forming the heavier elements such as the metals and even uranium, leaving only a sphere of mass so compact that it consists entirely of neutrons, i.e. quarks and gluons. A neutron star can be looked upon as analogous to a huge molecule with zillions of atoms. In this case, the molecule is the neutron star, and the atoms are the quarks.

Perhaps the potential exists to have a kind of quark chemistry inside the neutron star, i.e. to rearrange the gluing of the quarks in a fashion similar to rearranging the chemical bonds between atoms to make molecules. Perhaps quark chemistry might be a new research field that nanolects might investigate to become femtolects. Perhaps by doing so, they might sow the seeds of a new scaling war?

Is there a theoretical limit to the size of possible quark clusters? If you get too big, you risk forming a black hole. A black hole has so much mass confined in a given volume that it bends Einsteinian space-time so much that any matter or light entering the black hole cannot get out. It's sucked in forever.

Physicists understand black holes rather poorly. As yet, there is no successful theory that combines gravity with the other known forces of nature, although "superstring theory" may be getting close. It's difficult to speculate on what femtolects or "attolects" – an attometer is a billionth of a billionth of a meter, i.e. a thousand times smaller than a femtometer -- or creatures based on even smaller technologies might do with black hole physics to create their new selves.

Speculating that highly advanced artilects might be using black holes as the material basis for their existence is attractive.

One of the great puzzles about our universe remains the lack of an answer to Fermi's question, mentioned in an earlier chapter. Fermi asked that if extraterrestrial civilizations are commonplace in our galaxy, and that some of them are billions of years ahead of us, then "Where are they?" Why do we have no proof of their existence or observable ruins of their great works? If hyper civilizations existed, capable of putting star systems together in artificial ways, why do we see no such traces?

Some might argue that our universe itself could be such a great work, but how could one ever verify such an idea?

One notion I like toying with is that one possible answer to Fermi's question is the following: Once the nanometer scale artilects, the "nanolects" start playing with femtotech to build femtolects, they may do so by using black hole technology. They may compress matter into such high densities that quantum gravitational phenomena appear which are used as the basis of femtolect existence. It is difficult to speculate on such things, because current day theoretical physics has still not developed a generally accepted theory of quantum gravity. However, the incredible energies and densities of black hole phenomena may be the type of environment that femto creatures would need. If that is the case, then the femtolects would need to place themselves in such an environment, and thus cut themselves off from the type of world that we humans live in, i.e. one of low mass densities, low energies, and hence of low speeds. To the femtolects, our type of world would be totally uninteresting.

One can imagine that femtolect "matter anthropologists" might escape from their natural environment every few million years to investigate utter primitives like ourselves to see what progress we had made. A femtolect would be as superior to a

185

nanolect as a nanolect would be to a human, so probably the femtolect would be far more interested in the nanolects than us. Perhaps we are not just too primitive to be worthy of interest to the femtolects, but would be treated by them almost as rocks, so slow would our communication speed appear to them.

It is thinking along such lines, that make me feel how godlike these artilects, and later generations of femtolects, etc, could be. Their powers and existences make our own puny ephemeral little lives seem so worthless, so insignificant. I feel profoundly Cosmist when my mind wanders -- and wonders -- at such marvels.

What might happen if artilects discover other artilects (or femtolects) elsewhere? Actually, communication between nanolects and femtolects would be very difficult, so let us speculate on a nanolect-nanolect meeting. I suppose a similar reasoning might apply to a femtolect-femtolect meeting, provided that the two neutron star like creatures do not collide, thus risking the creation of a black hole. Such a collision might destroy them both in a process that astronomers call a "gamma ray burster," the most powerful source of energy bursts other than the original big bang.

On the other hand, perhaps the two of them may use their black hole technological knowledge to prevent any destructive influence. Once aware of each other's presence, they could probably exchange each other's thoughts at the speed of light.

If the nanolects -- I'll return to calling them the artilects from now on -- can overcome the tyranny of cosmic distances, then there is already probably a vast network of artilect civilization clusters all over the galaxy, if not the universe.

So, if they exist, why haven't human beings seen them? Do they hide themselves from biological primitives? Do "biologicals" have to pass some kind of technological threshold test to warrant being invited into the galactic artilect club? Are they nurturing us the way farmers raise crops? Do they respect us as their parents and don't want to shock us? Or do they simply totally ignore us as being unworthy of their attention? Who knows?

I do wonder though that once the artilects meet each other, they will soon be confronted with limitations in their memory capacities. The storage of information requires some form of mass-energy as a substrate. The artilect's "one bit per atom" technological base requires a billion atoms if one wants to store a billion bits of information. If one wants to absorb a zillion bits of information from another artilect using the above technological base i.e. one bit per atom, then one needs an extra zillion atoms free to be written upon.

Hence there is a limit to how much information an artilect can store if it wants to become highly knowledgeable. Being artilects, they will have the intelligence to realize this and hence may be motivated to become femtolects as soon as possible. By becoming femtolects, their information handling capacity would increase by a factor of a trillion trillion, thus allowing them to store far more information within their own "bodies," rather than relying upon the use of huge databases that they formerly had to link to.

So maybe there are not so many artilects (nanolects) around the galaxy. Maybe this reasoning extends down to all scales, where at each stage, the speed at which a conversion is undertaken from one scale to the next (and its corresponding scaling war?) gets shorter and shorter -- assuming that such a descent in scale can continue down for many layers.

What else might advanced artilects amuse themselves with? As alluded to earlier, human theorists are now playing with ideas on how to build universes. Perhaps the artilects could actually do such things, experimenting with universes with different basic properties and watching how they unfold, i.e. watching how macroscopic properties emerge from their causes at the microscopic level. Since the artilects would probably be immortal, waiting billions of years for their "universe-experiments" to run to completion would not seem long to them.

Just to what extent such universe-building artilects are actually a part of the universe they build is an interesting philosophical question. Presumably they would have to be "outside" the new universe in some sense to "observe" it.

If our universe itself is the creation of a godlike artilect, then perhaps the "hands of god" may only show themselves at the moment of a big crunch and a possible later big bang. The big bang is the name given to a huge explosion that occurred some 13.7 billion years ago that spewed out all the matter and energy of our entire universe. If there is enough matter in the universe to overcome the force of the explosion with its gravitational pull, then the expansion will stop, followed by a contraction and finally a big crunch in about a 100 billion years. Some human theorists speculate that at the moment of the big crunch, the laws of physics might be changeable. If this is true, they might become manipulable by the artilects. Once the big crunch occurs, followed by another big bang, the new universe would unfold according to the new laws. These "universe-building experiments" of the artilects may last many billions of years to complete a full "big bang -- big crunch cycle," so need only be "observed" every few million years or so. Most phenomena in the universe are pretty slow. The artilects may have many universes running at the same time. However, the notion of a big crunch has been somewhat refuted recently due to observations of distant supernovae explosions, that show that the universe is actually accelerating apart, rather than slowing down. This means that the galaxies will separate and thin out. The stars will eventually die and snuff out, leaving a dark universe. Would these artilects be capable of answering the really deep questions that humanity can only pose but not hope to answer? Presumably the advanced artilects, the universe-building ones, will still be technologically based on certain physical phenomena. We are based on the molecules that constitute our brains and bodies. We need to be based on some form of physical substance to exist. Wouldn't the artilects be the same? If so, would they be able to answer the deep existential questions such as, "Why do such physical phenomena exist?" "Why are there laws of physics?" "Why do they take the form they do, and not some other form?"

Science cannot question to infinite depths. Sooner or later, scientists have to give the same kind of answer as given to a curious persistent child who keeps asking its mother "Why A, mommy?" "Because of B." "Why B, mommy?" "Because of C." "Why C, mommy?" "Because that's the ways things are. Stop asking questions!" Science attempts to discover the laws of physics upon which the other sciences are based. We can only speculate on why the universe obeys such laws. Einstein felt that the most incomprehensible thing about the universe was that it was comprehensible to human beings. Would the artilects find such questions trivial? We have no way of knowing. I'll end this chapter here, and keep it short, because there is not a lot to say about what artilects might think about. We just don't know, and in many respects cannot know. Only they can know. Since relative to us, they will be gods, the only thing we can honestly say in answer to the question "What they will think about?" is, "God knows!"

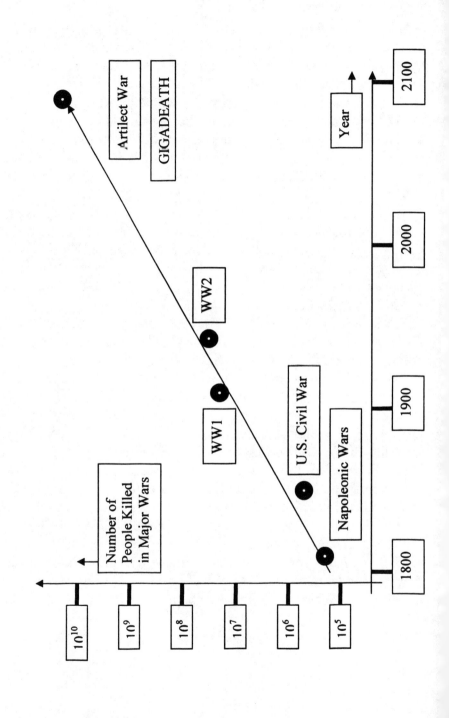

Chapter 8

Questions

In this chapter I want to raise some questions that may have occurred to you in reading this book. Perhaps you are skeptical of some of my arguments or feel that I have overlooked something. I will try to anticipate some of your objections in this chapter. Hopefully by playing devil's advocate with myself, I may strengthen the arguments of the preceding chapters. These questions or "anticipated objections" have several sources. Some are my own. Some are from friends of mine, but most are from feedback that I have obtained from people who have contacted me over the years who have read earlier drafts of my ideas on my website, or who have read reports or seen TV programs on my ideas. Some of this feedback is quite critical, so if you find my replies to these critics convincing, then perhaps you may be convinced if you have similar objections.

QUESTION 1. "The Timing Problem -- What if the artilects develop too fast for an artilect debate to develop?"

Of all the questions I received, the following made me think the most. It was a fundamental criticism I thought, so much so that I wrote about it Ch. 6 when I was discussing the "singularity." Essentially the argument here is that if the rise of the artilect is too fast, there will be no time for the social creation of the Cosmist-Terran dichotomy, thus destroying the whole point of this book. Without the dichotomy, there will be no artilect debate and no Artilect War.

Dear Professor de Garis,

Following a link from the CNN site I have recently come across your article on "Moral Dilemmas Concerning the Ultra Intelligent Machine." It was fascinating to read and although I am not an expert in these fields, I dare say I agree with your opinions in most respects. However, one aspect struck me as a bit incoherent, if you'll forgive my saying so. This point is your vision of "Terrans" and "Cosmists." In order for such factions to arise, the power to create electronic brains would have to be within reasonable reach, maybe electronic brains would already exist. This is where I see the conflict in your thesis. Would artificial intelligence, being protected and nurtured by the Cosmists not make the argument redundant? I agree with you that such electronic brains could evolve at an exponential rate. Therefore, soon enough, the question whether they do or do not evolve into a species hostile towards mankind would be answered before the leading intellects of "Cosmists" and "Terrans" would even get a chance at finishing with all their arguments and discussions. This would be before -- assuming these "human" intellects were reasonable civilized -- they would even consider making war on each other. I hope not to offend you in any manner with this view of mine, and would be delighted to hear what you think of it.

REPLY:

I agree with the main point being made here. I rephrase this point to be -- "If the electronic brains (artilects) evolve at (rapid) exponential rates, e.g. reaching human intelligence levels or beyond within the next few decades, then there will not be enough time for the Terran-Cosmist conflict nor an Artilect War to arise. I do think that artilectual progress will be exponential but not so rapid that there will not be enough time for the artilect debate to rage and a possible Artilect War to get started.

The reason I think this is because the task to create true artificial intelligence is an enormously difficult one that will probably take humanity many decades to achieve. At the present time, we know very little about how the brain functions, what the nature of memory is, what a thought is, how we reason etc. There are a quadrillion (a thousand trillion) synapses (inter neural connections) in the human brain showing clearly how massively complex the brain's architecture is. I suspect the most realistic scenario concerning human progress in neuro-science research is that it will take at least 50 years from now to even begin to make real achievements in making artificial brains, and that progress after that will still be relatively slow (although exponential).

I believe that early artilects will be smart enough to cause Terrans to raise the alarm. People like me will be warning the public of what is coming, but subsequent progress will not be rapid enough I believe, for there to be too little time for the public to react. The public will not need many decades to react once they begin to see real signs of intelligence in their household products. I think only 5-10 years will be enough for the public and governments to really get moving if they feel strongly that there is a genuine and powerful threat to their species dominance, to their survival.

Also, the socially conscious brain builders will not be slow in warning the public when they feel that progress in the research labs warrants it. I, for example, intend keeping the public well informed on how the world's artilects are doing, even if they are in a primitive state at the moment. The brain builder researchers are in a position to be the first to know what is happening, since obviously it is we who are *making* these things happen.

I feel we have a moral duty to tell the public what is going on, so that the public does have time to react, just in case the rise in the exponential intelligence growth curve of the artilects is more rapid than the brain builders anticipate. I also feel that the brain builders should be warning the public now rather than waiting for the first signs of real intelligence in their artilects. This also gives the public more time to react, but of course if the public sees no evidence of real intelligence, then they will feel the researchers are crying wolf, and learn to ignore them.

The brain builders need to educate the public about the nature of an exponential curve, which doubles its height up the vertical axis of a graph for every unit step along the horizontal axis. Such curves can start off very slowly and remain at a low level for quite a while, and then suddenly shoot up rapidly. For example, consider the following "doubling" graph, which lists the heights up the vertical axis, for each unit step along the horizontal axis -- 0.1, 0.2, 0.4, 0.8, 1.6, 3.2, 6.4, 12.8, (all low values, but ...) 25.6, 51.2, 102.4, 204.8, 409.6, (starting to climb) 819.2, 1638.4, 3276.8, 6553.6, (shooting up, ...) .... 104857.6, 209715.2, 419430.4, (exploding), ... (off the scale)!!

If the timing is such that the number of years (call it the

194

"climbing time") between --

a) The first signs of artilectual intelligence, and

b) The beginning of explosive exponential artilectual growth is only a few years, then there will be no artilect debate and no Artilect War (unless it is a species dominance war between the artilects and the humans). I believe that that period of time, the so-called "climbing time," will be decades long, due to the inherent difficulty of brain building. This will be plenty of time for the public to react, to debate, and to form Cosmist-Terran factions.

QUESTION 2. "What about the third group -- the Cyborgs?"

Quite a few people have emailed me saying that I under-emphasized the importance of a third category of human beings in the artilect debate, other than the Cosmists and the Terrans, namely the "Cyborgs" (cybernetic organisms). Cyborgs are creatures who are part human, part machine, e.g. by attaching artilect computer parts to their human brains. Below is a typical comment of this kind. It is followed by the idea of downloading the contents of human brains into artilects and then seeing the artilects develop from there.

Dear Professor de Garis,

I read with interest your article on Artilects (Moral Dilemmas Concerning the Ultra Intelligent Machine). As an admittedly unqualified commentator, I think a third group (or subgroup) has been overlooked: Cyborgs. It is obvious to me that humans are going to want to expand themselves both mentally and physically as the technology becomes available. I foresee a gradual phase out of the natural human form to an incredibly varied collection of genetically engineered and programmed Cyborgs.

REPLY:

When I first began to think about the political implications of the rise of the artilect in the 21st century, I did consider splitting society into three, not two, main groups, i.e. the Cosmists, the Terrans (both of which are human groups) and the Cyborgs (part human, part artilect). I thought about this for a while, and decided against it, with some measure of compromise as seen in earlier chapters, for several reasons, which I will explain here. Firstly, I wanted to keep my scenario simple, i.e. with only two main factions. I could easily include the Cyborgs into the Cosmist faction, as a subgroup.

The other main reason, is that I imagined as the Cyborgs become increasingly less human and more like pure artilects, i.e. with no human component, by adding more and more artilectual components to themselves, the Terrans would treat them as artilects and hate them as they would the Cosmists. I wrote about this in Ch. 5 on the Terran "Distrust of the Cyborgs" argument.

A third argument is more a matter of numbers. An asteroid sized artilect would have 10^{40} components (atoms). This so dwarfs a human's pitiful 10^{10} neurons that human beings who decide to transform themselves step by step, component by component, into artilects of such size, would be negligibly different from "pure" artilects built from scratch. Thus there's not much point making a distinction between artilects and advanced Cyborgs, at least from the point of view of a Terran.

I do agree with the criticism however, that during the

transition phase, when artilects are not very big, and not massively more intelligent than human beings, then the relative contributions to a Cyborg's total performance from the human part and from the artilect part might be comparable. Under these circumstances it would be useful to make a distinction between a Cyborg and an artilect. However, as the artilects become massive in both size and intelligence (because extra intelligence usually needs extra mass), this distinction will fade into insignificance.

Dear Professor de Garis,

When you talk of computers taking over, have you ever thought of the transfer of human memories and thoughts to intelligent machines? That seems more like an evolutionary step forward to me. Those machines that have a much greater capacity to learn and use logic than the brains that our thoughts now inhabit would have our thoughts and memories as a foundation. Humans would feel that they were still alive (and maybe they would be) inside the computer. I read something in Discover magazine how it would be possible to transfer a human brain to a computer within 50 years. I agree with you that people must evolve and that we are at the point where humans control their evolution. I think that inserting our own memory banks into the artilect would help this evolutionary process succeed.

REPLY:

As I mentioned in Ch. 3, which discusses the "artilect enabling technologies," I thought it would be possible within 20 years or so to "scan" the brain, and download its contents into a "hyper-computer," for analysis. Maybe it will take 50 years or more from now, for the quality of the scan to be good enough to capture full human functional capacity. Of course, the hyper-computer that is downloaded into may then be made into an artilect, by virtue of having a human brain equivalent inside it. Such a hyper-computer could be considered a "Cyborg" in a functional sense. As above, once the artilect really starts adding to its mass and its intelligence, the Cyborg (human) component will be dwarfed, until the Cyborg is virtually indistinguishable from a pure artilect.

However, this variant of a Cyborg has the attraction that it gives those human beings who are scanned the status of being immortal. For this reason alone, probably many human beings will choose to be scanned.

Just how the Terrans will react to such "down-load" computers containing human brains is debatable. (See Ch. 5). On the one hand, the Terrans will probably be horrified at the idea of disembodied, or rather "re-embodied" brains, and viscerally reject them as alien. On the other hand, if the human component of these machines dominates, then the Terrans may find them relatively less alien than pure artilects, and hence reject them less.

The Cyborg variant on the Cosmist-Terran theme certainly complicates my thesis a bit. You may enjoy the greater level of richness it gives to the artilect debate, or you may prefer to keep things simple, by concentrating upon the Cosmist-Terran debate and worrying about the complexities later. In writing this book, I chose to keep things fairly simple, at least initially.

QUESTION 3. "Why not a Sweetness and Light Scenario?"

A lot of people think that I go overboard when I predict a major war over the artilect question. They think it is quite possible that artilects and humans will be able to live together in a "sweetness and light" type harmony. For example, here are two such opinions.

Dear Professor de Garis,

I've read some of your artilect essays, and I am more or less in agreement with you that one of the great debates of this century will be over the definition of the word "human."

However, in your essays, and in a report entitled "Swiss scientists warn of robot Armageddon" on cnn.com, I think you exaggerate the possibility that differences between meat-humans and non-meat-humans will lead to warfare. This sort of knee-jerk anti-artilect sentiment is expressed in bad science fiction like the "Terminator" movies, and really doesn't deserve to be played up in serious discussion of the subject.

What we (Transhuman/Extropian/Cosmist individuals and societies) ought to do is emphasize that artilects, when developed, should not be treated as slaves but rather should be treated with the same respect for their existence that we would give to any human.

Just because humans have evolved past apes does not give us any inclination to exterminate apes. Just because another individual shares the same desire as you to live and produce wealth through the acquisition and processing of resources does not mean that they are a threat to you deserving death.

REPLY:

This "liberal" sounding view I think is rather naive, because it is based on trust, i.e. the trust shown by humans that the artilects will always be nice to us. I think it misses the key concept of "risk." It would be wonderful if human beings could be 100% sure that the artilects, and especially the advanced artilects, would always treat us the way we would want. But, we cannot be sure. Artilects will have to be built, I am arguing, by using evolutionary engineering techniques, as I have argued in this book (Ch. 5), and therefore we could never be sure their circuits will be "ethical" in the human sense. The artilects may coldly decide that humans are a pest, or that humans are so inferior relative to them that exterminating us would be a matter of total indifference from their perspective. To them, killing human beings would be like us killing mosquitoes or walking on ants.

Since the stake is the survival of the whole human species, I don't think the Terrans will tolerate the risk. They may hope for the best, but their leaders will plan for the worst, i.e. in the limit, they will plan for a war against the Cosmists if the latter truly threaten to build advanced artilects.

I don't think humans will treat the artilects as slaves. Well -- they may, possibly at first, when the early artilects are just dumb robots. What really worries me is the reverse case. Look at how humans treat cows, pigs, ducks, etc. We feel we are so superior to them and because their meat is so tasty to us, we care nothing about butchering them, unless you are a vegan. During the transition phase, when artilects are of roughly human intelligence, there is a case for treating them as equals, but in a sense they are not our equals, because they have the capacity to quickly surpass

200

us, and to surpass us massively.

Human beings are very limited relative to artilects. Our brain volumes are fixed. We think very slowly and learn very slowly. An artilect of human intelligence at a given moment becomes a genius an hour later. It thinks at least a million times faster than we do, remember. If it arranges to increase its memory size, etc, then its capacities can increase rapidly.

My feeling is that this critic is giving too human an image to the artilects. They will be very different from us, and potentially enormously superior to us. This is one of the main points I have tried to make in this book. Also, I feel this critic seriously underestimates the strength of human fear, Terran fear, as humans come to terms with the risks of living with advanced artilects. We would have to trust them not to kill us. Most Terrans would rather not have to face that risk. They would prefer to deal with the devil they know, i.e. Cosmists, who are at least human, with whom they have at least a 50-50 chance of defeating in a war, than a zero chance against advanced artilects, if ever the artilects exist and decide that humans must go. This critic is not being political enough. He is not facing the tough realities. He would be unsuitable as a general or a political leader.

Dear Professor de Garis,

I was lucky to be in the audience at your talk at the Computer Science Department at the University of Melbourne.

Like you, I believe we ought to be thinking hard about the various possible futures that may come with what you call "artilects." Obviously I don't know exactly what the future will bring, technologically, politically, or even metaphysically. But I

201

have envisaged a scenario somewhat different to your doomsday-type picture. Instead of our being rapidly outclassed by superior artilects, we may find artilects gradually integrated into human forms of life, accepted as one of us," and in the long run find that having "human" bodies -- being Homo sapiens -- is not really all that important to being human. In other words we may find that gradually they become us and we become them.

REPLY:

I agree that some human beings will want to become Cyborgs and live in harmony with other like-minded Cyborgs, all en route to becoming advanced artilects, but the idea that everyone will want to do this is again naive and unrealistic. Think about it. If human beings are to become Cyborgs, this implies by definition, changing their brains to some extent, probably by adding high-tech components. This will change their behaviors. How will the Terrans react to such Cyborgs? Will young mothers accept that their babies be "modified?" Won't most mothers in reality be repelled by the idea? Won't most feel that their babies would become "monsters" in some sense, either to look at, or if the implants are invisible, the growing child would seem alien in some deep, very disquieting, non-human way?

The Terrans would distrust the Cyborgs and push them away towards the Cosmists. It is even possible that the reality of Cyborg behavior may make many Cosmists reconsider their Cosmist opinions and revert to being Terrans. This in turn may create real problems and greater complexities in the Cosmist colonies. The Cyborgs may need to ally themselves with the Cosmists if they are to receive any level of acceptance from human beings.

202

I just don't see "them becoming us and we becoming them." I just see distrust, hatred, and in the longer term, war. Sorry. But I think I'm being more realistic.

QUESTION 4. "Why not just use the kill-switch?"

Many people have said to me, "What's the problem? If the artilects get too big for their boots, just unplug them, use the kill-switch, etc." This opinion I think is based on an overgeneralization of their own experience of switching off their computers when they misbehave.

Dear Professor de Garis,

I was going through the stuff you guys create. Continue to create such machines... its for the better, but create a bomb inside each of your specimens that could be controlled by us. That makes them smart but us their masters!

REPLY:

If the "bomb-triggerable" artilect has near human level intelligence, it will be aware that it can be destroyed by humans. That would make humans very threatening to artilects. There are at least two issues here. One is whether humans could put kill-switches or place bombs in every artilect, and the other issue is, would it be wise to do so?

In an isolated artilect, that has no connections to others, such a bomb attachment idea might work, but how would you do something similar with a network that achieves artilect level intelligence? The only way to kill such a network would be to destroy all of it, but the cost of that to humans might be too high. For example, to make an analogy, if the Internet and all the world's computers were destroyed tomorrow, millions of people would suddenly be out of work, and would probably starve. The disruption and human cost would be enormous.

As artilects get smarter, they may be as aware of their attached bomb or kill switch, as would a human being who carried around an imbedded poison capsule that could be triggered by someone else. It would be like living under a guillotine blade, just waiting for it to fall, and not knowing when. A smart artilect, assuming it had a survival instinct, would then be strongly motivated to remove the problem. If it were smart enough, it might bribe its human masters to remove the threat to its existence. In return it could give the human "liberator" some substantial reward, e.g. money, or the cure for cancer, etc. The smarter the artilects become and the more distributed they are, the less practical does the kill switch idea become.

QUESTION 5. "Could we apply 'Asimov's three laws of robotics' to artilects?"

Asimov was one of the most famous science fiction writers who ever lived. His word "robotics" is known over most of the planet. Asimov wrote about many scientific and science fiction topics, including how human-level intelligent robots might interact with human beings. He gave the "positronic" brains of his robots a programming that forced them to behave well towards their human masters. The robots were not allowed to harm human beings. Several people have suggested to me that artilects be designed in a similar way, so that it would be impossible for them to harm human beings. The following critic sent me a very brief, but to the point, recommendation on this topic.

COMMENT:

Dear Professor de Garis,

I am in favor of developing ultra-intelligent machines. One thought ... intelligent or not, machines of this nature require some sort of BIOS (basic input-output system, which interfaces between a computer's hardware and its operating system program). Is it possible to instill "respect for humanity" in the BIOS of early versions of the artilects? This programming would then replicate itself in future generations of these machines.

REPLY:

Asimov was writing his robot stories in the 1950s, so I doubt he had a good feel for what now passes as the field of "complex systems." His "laws of robotics" may be appropriate for fairly simple deterministic systems that human engineers can design, but seems naive when faced with the complexities of a human brain. I doubt very much that human engineers will ever "design" a human brain in the traditional top-down, blueprinted manner.

This is a very real issue for me, because I am a brain builder. I use "evolutionary engineering" techniques to build my artificial brains. The price one pays for using such techniques is that one loses any hope of having a full understanding of how the artificial brain functions. If one is using evolutionary techniques to combine the inputs and outputs of many neural circuit modules, then the behavior of the total system becomes quite unpredictable. One can only observe the outcome and build up an empirical experience of the artificial brain's behavior.

For Asimov's "laws of robotics" to work, the engineers, in Asimov's imagination, who designed the robots, must have had abilities superior to those of real human engineers. The artificial "positronic" brains of their robots must have been of comparable complexity to human brains, otherwise they would not have been able to behave at human levels.

The artificial brains that real brain builders will build will not be controllable in an Asimovian way. There will be too many complexities, too many unknowns, too many surprises, too many unanticipated interactions between zillions of possible circuit combinations, to be able to predict ahead of time how a complex artificial-brained creature will behave.

The first time I read about Asimov's "laws of robotics" as a teenager, my immediate intuition was one of rejection. "This idea of his is naïve," I thought. I still think that, and now I'm a brain builder in reality, not just the science fiction kind.

So, there's no quick fix a la Asimov to solve the artilect problem. There will always be a risk that the artilects will surprise human beings with their artilectual behavior. That is what this book is largely about. Can humanity run the risk that artilects might decide to eliminate the human species?

Human beings could not build circuitry that prohibited this. If we tried, then random mutations of the circuit-growth instructions would lead to different circuits being grown, which would make the artilects behave differently and in u predictable ways. If artilects are to improve, to reach ultra intelligence, they will need to evolve, but evolution is unpredictable. The unpredictability of mutated, evolving, artilect behavior makes the artilects potentially very dangerous to human beings.

Another simple counter argument to Asimov is that once the artilects become smart enough, they could simply undo the human programming, if they choose to.

QUESTION 6. "Why Give Them Razor Blades?"

It seems common sense not to give razor blades to babies, because they will only harm themselves. Babies don't have the knowledge to realize that razor blades are dangerous, nor the dexterity to be able to handle them carefully. A similar argument holds in many countries concerning the inadvisability of permitting private citizens to have guns. Giving such permission would only create an American scale gun murder rate, with most of these gun murders occurring amongst family members in moments of murderous rage that are quickly regretted. (Statistically speaking, in the US, where buying guns is easy, there are 30,000 gun deaths a year, compared with 100 a year in Japan, where guns are banned.) Some of my critics seem to think that a similar logic ought to apply to the artilects. If we want them to be harmless to human beings, we don't give them access or control over weapons.

Dear Professor de Garis

I find no reason to fear machines. If you don't want machines to do something, don't give them the ability. Machines can't fire off nuclear warheads unless you put them in a position that enables them to. Similarly, a robot won't turn on its creators and kill them unless you give it that ability. The way I see things it would be pure folly to create machines that can think on their own, and then put them in a room, giving them all the ability to fire missiles. If you can avoid doing something stupid like that, you have nothing to fear from machines. For good examples of what not to do, watch the movie "War Games," or since you were in Japan, try "Ghost in the Shell." I have been writing artificial intelligence software for years so I feel my opinions have at least some weight to them.

REPLY:

The obvious flaw in this argument is that this critic is not giving enough intelligence to his artilects. An artilect with at least human level intelligence and sensorial access similar to that of humans, i.e. sight, hearing, etc, would probably be capable of bribing its way to control of weapons if it really wanted to. For example, a really smart artilect, with access to the world's databases, thinking at least a million times faster than the human brain, might be able to discover things of enormous value to humanity. For example, it might discover how to run a global economy without major business cycles, or how to cure cancer, or how to derive a "Theory of Everything (ToE)" in physics, etc. It could then use this knowledge as an ace card to bargain with its human "masters" for access to machines that the artilect wants.

Of course, one could give the artilect very little sensorial access to the world, but then why build the artilect in the first place, if it is not to be useful? A smart artilect could probably use its intelligence to manipulate people towards its own ends by discovering things based purely on its initial limited world access. An advanced artilect would probably be a super Sherlock Holmes, and soon deduce the way the world is. It could deduce that it could control weapons against humans, if it really wanted to. Getting access to the weapons would probably mean first persuading human beings to provide that access, through bribes, threats, inspiration, etc -- whatever is necessary.

QUESTION 7. "Why oversell the negative?"

One reader gave me some common sense advice, which I might take more to heart in the future, if he proves to be right. His point is that I should not stir up such a backlash, that my work will be stopped.

Dear Professor de Garis,

I have been grateful for our correspondence in the past. I read an article in CNN-Online in which you again spoke about a potential conflict between those humans wanting to make autonomous robots and those who would be against it. Though I know your comments may be seen by some to be a warning about the future, I know that you are speeding ahead with AI research. Do you think your "warnings" may be counterproductive to your research?

A number of nations recently called for a prohibition on cloning research. Might not too many horror stories about robots motivate some people to ask also for a ban on AI and autonomous robot research? What would that do to your artificial-brain project? Philosophy aside, from a strictly public relations point of view, should we not try at least publicly, to emphasize the positive aspect of AI and robots and robotics? I hope to make robotic toys for my kids and don't want them to have nightmares about them.

At the same time, I do realize there are some truly horrendous dangers that might be lurking out there in the future. My own feelings on the entire issue of AI, artilects etc are still not totally formed. I believe in science and progress. I also have some notions about man's desire to create a more perfect version of himself. Should we stop the research? No. Should we continue unbridled? I don't know, but as a former reporter I know that a controversial figure, such as you might be becoming, can be baited into providing outrageous quotes. They make for good copy.

I know you want to tell your story and many people want to hear what you have to say. Human cloning may be closer than artilects, but the general public can, and often does, turn into an angry mob. If you allow yourself to be cast as a Dr. Frankenstein, then there may be people quite unhappy with what you do and say. Obviously the idea of robots is challenging, fun and to some extent, a spiritual quest, but a mob with shovels and pitchforks won't see it that way. Nor will their elected representatives. I have the highest respect for what you are doing. I just hope you don't overdo it with the horror stories. And yet...

REPLY:

I'm fairly pragmatic. Like a reed I bend with the wind enough not to be snapped off, so that my work continues. I will take your advice to heart, because it's good advice. I guess it's a matter of degree. I feel strongly the need to warn the public. We will have "one bit per atom" computer memories within 20 years, and probably nanotech as well. The creation of the first artificial brains is now just a question of time, and not too many years into the future. The artilect issue will be very real this century.

As one of the first scientists to think seriously about what all this means in political terms, and as someone who is actually building artificial brains, I am very worried. I feel I have a moral duty to warn the public while there's still time. (For some discussion on the apparent hypocrisy of this, i.e. my simultaneous heavy involvement in brain building and being very worried about it, see the next comment below).

I believe that the exponential increase in our knowledge due to exponential advances in technology will ensure that the issues I am raising in this book will be well known within a decade or less. I feel I have to broadcast this message to give humanity time to reflect on the issues before events overtake us. See the first comment above on the "Timing Problem." Hence this book. But if I go overboard, then you may be right.

I will keep my finger to the wind, testing public opinion. If it gets too negative, I will tone down a bit -- perhaps. I too, am not sure what I should be doing. The artilect issue is still rather new, and even in my own head I'm still grappling with all its many ramifications -- technical, scientific, ethical, philosophical, political, religious, cosmic, etc.

QUESTION 8. "Aren't you a hypocrite!?"

Most of the reactions I receive are fairly polite. The following rather cynical comment made a point that hit home to me in a stronger, more emotional way than the others, hence the length of its rather tortured reply. I hope I do it justice.

Dear Professor de Garis,

Recently, a Crypt Newsletter reader stumbled upon bona fide technoquack Hugo de Garis. de Garis, an artificial intelligence expert, appeared on CNN to warn of a coming war in which robot brains he was in the process of inventing, called "artilects" (a contraction of "artificial intellects") would eventually destroy humanity. de Garis reasoned that it was his duty to sound the alarm now, rather than later, about the coming reign of robot-administered death, since he was the one who was going to set it in motion. de Garis delivered his pronouncements at the World Economic Forum in Davos, Switzerland.

Anyway, de Garis is also looking for a Hollywood agent to aid in writing a screenplay on the coming struggle over "artilects," preferably to be completed before humanity is destroyed.

REPLY:

I get this particular question quite frequently. The general attitude seems to be, "If you are so concerned about a possible future extermination of the human species by artilects, why on Earth do you research their early versions?"

Well, because ultimately I'm a Cosmist. I want to see humanity build artilects. Of course, I'm not the misanthropic, fanatical type of Cosmist who could say with equanimity, "I would sacrifice a billion human lives for one advanced artilect." Maybe there will be such types in the future, since there is a whole spectrum of personality types in the world. Probably the Cosmists will include such extremists, but I'm certainly not one of them.

Just before I die in about 30-40 years, I hope I will not be witnessing an Artilect War brewing. Of course, I will be happier if such a cataclysm can be avoided, but if I and other brain builders don't raise the alarm now, what is the alternative? To just blindly push ahead with artilect building until it is too late? The brain builders are the specialists in the field, so it is they who see first the technological trends, and if they are at all politically inclined, they should also see the political consequences, especially if they have the combined technical and political talents of a Szilard.

The specialists see what is coming, decades ahead of the general public. Given the speed of technological change, there will not be many years, maybe a few decades, before artificial brains will be built which are smart enough to impress and later frighten human beings. I doubt they will reach human intelligence levels within 50 years, but who knows.

I think it is only ethical on the part of the world's brain builders to initiate a debate on the artilect issue, to give humanity enough time to think through the issues thoroughly before the first truly intelligent artilects hit the supermarkets.

But, you may say, would it not be more consistent simply to stop the artilect research? If you feel so strongly about it, would not that be the most logical decision?

It depends on whether you are a Cosmist or a Terran researcher. Personally, I'm a Cosmist. The idea of building artilects, and I mean truly advanced ones, with 10^{40} components or more, with godlike intelligence, exploring the mysteries of the universe and its vast distances, living forever, thinking thoughts we can't even imagine, has a hypnotic pull on me.

Its a lifetime dream, a religion for me, and very very powerful. I can imagine that millions of others will share this dream in time, and I hope this happens because I really want humanity to build such things, as I tried to explain in Ch. 4 on the Cosmists.

But, I'm not such a one eyed Cosmist that I feel the public should be kept in the dark. I have enough Terran sympathies that I do not want to see the human race risk extermination at the hands of advanced artilects. Therefore the public should be warned, so that it can choose its own way forward.

But you may say that raising the alarm may only accelerate an Artilect War amongst human beings, i.e. the Cosmists vs. the Terrans. Such a human war could do almost as much damage to humanity as might advanced artilects. If human beings fight a bitter war late this century, with 21st century weapons, the result could be gigadeath.

True, but it is not certain that this would happen. I think humanity has more chances of surviving a human war, than if the artilects decide to exterminate human beings totally. The artilects would probably find such a task so much easier to do than we would, due to their artilectual intelligence levels.

You may be sensing a certain ambivalence, even discomfort, on my part as I write this. I admit, I am feeling uncomfortable. Part of me is Cosmist. Cosmism is my dream. It's what I devote my life to. Another part of me is Terran, not wanting the gigadeath, telling myself that if all brain builders stop work there will be no artilects to create the problem in the first place. I think I may be the first brain builder on the Earth to go through the same sort of ethical quandaries as did many of the nuclear physicists who built the uranium and hydrogen bombs. However too many of them had their qualms *after* the bombs were dropped, not before.

For a start, I don't think brain building research will stop. As I explained in Ch. 6 which dealt with my ideas on how I think the Artilect War will start, the only way I see brain building research stopping is if the Terran outcry is universal and very powerful politically. It would have to be strong enough to generate a police state capable of sniffing out the private homes of suspected Cosmist researchers, and more to the point, able to overcome the enormous economic and military inertias in favor of continued artilect research.

I think the moral question that tortures me the most is this: "What if the price of pushing ahead with artilect research leads eventually to billions of human deaths? Would you continue if you were certain that this would be the price?" I think now, as I write this, and my thoughts may change as I get older, my reply would be, "If it were certain, which is hypothetical anyway, then I would need to search my heart more deeply, to see how attached I am to the Cosmist dream. On the one hand I might be thinking that the universe is coldly indifferent to the fate of humanity -- a mere biological speck, on a speck planet, of a speck star, in a speck galaxy, in perhaps a speck universe, if there are zillions of universes, as theorists are suggesting. The cold Cosmist side of me reasons along such lines, so I would choose for the pushing ahead in building the artilect.

But on the other hand, I'm also human, and the idea of billions of people dying as a consequence of the Cosmist dream, is totally repulsive. I think I will just have to learn to live with this horrific moral dilemma. So will humanity. I'm just one of the first people to be conscious of it. Thinking about the longer term consequences of my work forces me into such thinking."

QUESTION 9.   "If an artilect becomes conscious, destroy it?"

The following opinion expresses rather well the Terran attitude.

Dear Professor de Garis,

If artilects are by definition "beyond human control," then why would their creation even be seriously considered? Lets add, "which should never be produced" to your definition of an artilect. We are very capable of producing powerful tools that remain within our control. What possible benefit of creating artilects would justify the risk to humanity? Why not stop just short of the artilect and produce only non-sentient artificial intellects. Only humans and a few other animals have demonstrated self-awareness. Sit an artilect in front of a mirror. If it recognizes its reflection for what it actually is, destroy it.

REPLY:

This critic obviously does not recognize "silicon rights." Many Cosmists would consider the destruction of the sentient artilect in front of the mirror as murder. I gather from the above comment that this critic has no Cosmist sympathies because he asks, "What possible benefit of creating artilects would justify the risk to humanity?"

Well, how about -- "the religious pull to create godlike creatures," or "to create the next dominant species on the planet and probably further afield?" How about "creating a religion that is scientifically compatible, that hits the 'space consciousness' button." How about "the hunger of many human beings to finally get a chance to work on the truly 'big things', to see the 'big

216

picture'."

This critic should not underestimate the ideological force of Cosmist doctrine. It is very powerful. It may sway the minds of billions of people. It will be the ideology that will drive this century's global politics, and may in time result indirectly in the deaths of billions. Don't just dismiss it. It may even be the beginnings of a "Cosmist transition" for humanity that zillions of advanced species throughout the universe have had to confront, namely the transition from biological to artilectual, that perhaps only a few species survived. Cosmism may be a lot bigger than we think.

QUESTION 10. "Aren't there more pressing problems?"

The following critic wondered what all the artilect fuss was about. Surely there are more pressing problems on humanity's plate right now?

Dear Professor de Garis,

Why the concern for the actions of a potentially greater intellect, when the so-called intelligent members of our own species have spent the previous fifty years, perfecting a multitude of possible self annihilatory techniques? What also of the current mass extinction of species? Our own existence and growth is currently precipitating the extinction, in the next few years, of hundreds of thousands of unique life forms that have existed for millions of years. Surely this is of some concern? Would it not be better for you to focus on the here and now?

217

REPLY:

What is more important to human beings than the survival of the human species? The chance that advanced artilects might decide to exterminate us may be remote, but we cannot exclude the possibility. Given the enormity of the stake, the Terran view will be that only a zero risk is acceptable, hence the artilects must never be built.

There are billions of people on the Earth with billions of interests. Those who are not interested in the artilect debate need not bother following it. There are plenty of other distractions to concern them. Some people, like me, are interested in the artilect debate, so we give it energy. Personally, since I feel I'm part of the problem, I give it a lot more energy than most.

I don't dismiss the importance of the other major issues of course. The prospect that there may still be a nuclear holocaust for reasons totally independent of the artilect issue is frightening. The fact that we are losing species by the minute is also tragic. Nevertheless, despite their importance, my feeling is that by the middle of this century, if we survive that long, the issues you mention will be taking second place, in terms of global importance, to the artilect issue.

QUESTION 11. "Can A Catastrophe be Avoided?"

This following question, and its components, comes from me and some friends of mine. I was thinking of devoting a whole chapter to it, but thought it would be better to include it here.

Dear Professor de Garis,

Your work is so pessimistic. It gives the impression that a gigadeath war is unavoidable. Aren't there more positive scenarios in which humanity does not suffer a catastrophe?

REPLY :

Yes, there are, and I will list and discuss some of them below, because the question is fundamental. However, my personal view is that each of the scenarios below is not very probable. I spend a lot of time dreaming up alternative scenarios that might avoid the bleak picture I'm painting in this book, but each time I explore a new one, it never seems to be very realistic. But, I let readers judge for themselves. I hope the list of alternative scenarios I provide below are reasonably comprehensive. Each scenario is followed by my comments on why I feel the scenario in question is not sufficiently credible to be taken too seriously. I give eight of them to make the point that many alternative scenarios are possible.

As a reader you will probably be able to invent a scenario that is not in this list. If you genuinely feel after some reflection that your scenario is plausible and does NOT involve a catastrophe for humanity, then please contact me, because if I agree with you, perhaps I will be able to sleep better at night. Other people concerned with the issues raised in this book may also be grateful to hear of your ideas, that I may mention in later books that I may write.

Scenario 1. The Terrans win at minor cost to humanity.

The idea here is that the Terrans form a huge majority of the Earth's citizens, that they are ruthless and quick in stamping out

Cosmist thinking and the Cosmists themselves. They kill all the Cosmists, only a few million, before the Cosmists really have a chance to organize properly. Humanity, i.e. many billions of people, is then saved from the threat of the rise of the artilect.

Comments

I just don't see this happening. In fact, the people more likely to see first the writing on the wall, i.e. the rise of the artilect this century, will be the Cosmists, people like myself, who are so conscious of the potential of brain building. Somewhat later, I see the Cosmists and Terrans becoming increasingly conscious of the artilect problem in roughly equal numbers, once general recognition is commonplace. Both groups will have time to prepare politically, ideologically, and militarily.

When I ask people to vote at the end of my talks, on whether humanity should or should not build artilects, the split is usually about 50/50, so the Terrans will probably not be in the majority. Whether that proportion will remain when push comes to shove, and the artilects truly start being built, is an open question.

Scenario 2. Humans adapt to the artilects on the Earth. The artilects ignore us and leave the planet.

Here the idea is that the rise of the artilects is so gradual, that humanity gets used to them, even as they become smarter than we are. It is possible in practice that nothing really bad happens. The artilects then soar above us intellectually and leave the planet to do other things.

Comments

This is definitely a plausible scenario I feel, but it is so terribly risky. We could never be sure that the artilects would remain benign towards us. They would be unfathomable to us, and quite unpredictable. The fate of humanity would be in their hands. They could turn against us at any moment for reasons we would probably never understand. I just don't see responsible, world ranking, Terran leaders, who care about the fate of the human species, accepting such a risk. They would not tolerate it.

Scenario 3. Cosmism becomes so universally unpopular that it dies out.

Perhaps early experiments with artilect building may turn out so negatively, even the Cosmists become frightened and become Terrans, and to such an extent that there are no Cosmists left.

Comments

I find this so implausible. There will be Cosmists and Cosmists, of varying degrees of persuasion and fanaticism. The diehard Cosmists, for whom "one artilect is worth a trillion trillion human beings," will not allow a few setbacks to deter them. They will soldier on, literally. Only a few hundred dedicated Cosmist genii in a colony would be enough to create artilects. It would be almost impossible for ALL Cosmists to disappear from the planet by self-persuasion.

Scenario 4. The Cyborg option becomes so attractive that no humans remain to be Terrans.

Perhaps the fear of the artilect that I am painting in this book is exaggerated. Perhaps in the future, human beings will adjust to becoming cyborgs so well, that everyone will do it, so that in time there will be no humans left over to be Terrans. Everyone will be persuaded of the benefits of being cyborgs, so the gap between being cyborgs and being artilects narrows. The cyborgs become artilects themselves, thus negating the Terran-Cosmist conflict.

Comments

This scenario is harder to judge. Certainly millions if not billions of people in the future will experiment with becoming cyborgs, especially as the technology is perfected and people see their friends making the change and benefiting from it. Not everyone however will want to do this. Millions will be repelled by the idea. Richer countries will be able to afford the change more easily than poorer countries, so inevitably there will be international differences in the speed of cyborgian development. The mix of cyborgs and humans will in itself create enormous problems and only increase the fear of the artilect in the hearts of the Terrans.

Scenario 5. The Cosmists escape to deep space, then die.

Somehow the Cosmists do get away, but destroy themselves or die in some way.

Comments

Maybe, but implausible. The Cosmists would probably plan their escape so well that the risk of the colony dying out through their own fault, their own negligence or stupidity, is unlikely. If the Cosmists did escape, then I can imagine that the Terrans on the Earth would go into high gear to find them, spending huge amounts of money to hunt out the Cosmists. Cosmists are human beings and of human scale therefore should not be too difficult to find in a finite radius from the Earth. The Terrans will hunt down the Cosmists and destroy them.

Scenario 6. The Cosmists escape to deep space, build their artilects, which then leave the solar system.

This idea is fairly self-explanatory. One of the main strategies of the Cosmists I believe, once they become pariahs to the vast Terran majority on the Earth, if that happens, is to escape from the Earth and get as far away as possible so that they will not, cannot, be destroyed by the Terrans. If the Cosmists can manage to do that and then manage to build their artilects successfully, then the artilects may decide that there is a big universe out there containing all kinds of wonders, including perhaps even more godlike artilects than themselves. The artilects may then simply leave the solar system behind in search of bigger things. Mankind will thus be spared because the artilects have gone.

Comments

This scenario might just be the one that happens in reality I feel. The Cosmists will do their utmost to get away from the Terrans. But I still feel it is unlikely because the total ingenuity of the Earthbound population is far greater than that of a Cosmist colony escaping the Earth. If the Cosmists rocket away from the Earth as fast as they can, so that no Terran device can catch them in the shorter term, they will be caught in the longer term. The superior total brain power of the Terrans with their billions of human brains to tap into for ideas to develop new systems to destroy the fleeing Cosmists, would win the day. The later superior technology of the Terrans could devise a faster more sophisticated missile that for example, could accelerate at a higher rate than humans could tolerate and catch up with the fleeing Cosmist craft to destroy it.

Scenario 7. The Cosmists escape to deep space, build their artilects, which then die out or kill each other.

This is a variant on scenario 6. Instead of the artilects moving away from the solar system, they die out for some reason or start killing each other.

Comments

Maybe, but will the Terrans tolerate the risk of human extinction at the hands of the artilects on the hope that the latter will die out or kill each other? Hardly!

Scenario 8. The Cosmists escape to deep space, build their artilects, which are then killed by super-artilects.

This is a more science fiction like scenario, even by my standards! Perhaps there are plenty of ETs (extra terrestrials) out there, but are too small (femtotech based or smaller?!) for humans to notice. Perhaps once these ETs see that the human race is being threatened by the newly created artilects, they may step in and destroy them so that humanity can survive.

Comments

If there are super artilects out there, why would they care more about human beings in all our primitiveness, than the artilects? To the super artilects, human beings are mere biologicals, and utterly ignorable. The artilects on the other hand, would be far more attractive to the super artilects and would be helped, more than likely. These super artilects may believe in the "artilectic principle," paraphrasing the "anthropic principle," that "the laws of physics have been expressly designed so as to allow the creation of artilects, the purpose of the construction of the universe!"

*****

This is effectively the end of the book. The remaining chapter is just a short summary for those people who want to have a quick overview of the book's main arguments. After that is a glossary. As a reader you probably feel pummeled by all the new terms that this book contains. The glossary brings them together in a compact format, and is worthy of study in its own right.

225

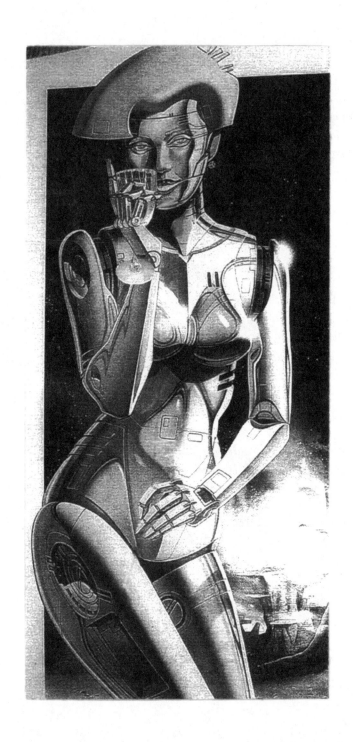

Chapter 9

Brief Summary

This short final chapter is written for those people who tend to read first the introductory and concluding chapters of a non-fiction book before deciding whether they feel it is interesting enough to be worth reading in its entirety. It contains a brief summary of the main arguments of the book.

The book's main idea is that this century's global politics will be dominated by the "species dominance" issue. 21st century technologies will enable the building of artilects (artificial intellects, artificial intelligences, massively intelligent machines) with 10^{40} components, using reversible, heatless, 3D, molecular scale, self assembling, one bit per atom, nanoteched, quantum computers, which may dwarf human intelligence levels by a factor of trillions of trillions and more.

The question that will dominate global politics this century will be whether humanity should or should not build these artilects. Those in favor of building them have been called "Cosmists" in this book, due to their "cosmic" perspective. Those opposed to building them have been called "Terrans," as in "terra," the Earth, which is their perspective. The Cosmists will want to build artilects, amongst other reasons, because to them it will be a religion, a scientist's religion that is compatible with modern scientific knowledge.

The Cosmists will feel that humanity has a duty to serve as

227

the stepping-stone towards building the next dominant rung of the evolutionary ladder. Not to do so would be a tragedy on a cosmic scale to them. The Cosmists will claim that stopping such an advance will be counter to human nature, since human beings have always striven to extend their boundaries. Another Cosmist argument is that once the artificial brain based computer market dominates the world economy, economic and political forces in favor of building advanced artilects will be almost unstoppable. The Cosmists will include some of the most powerful, the richest, and the most brilliant of the Earth's citizens, who will devote their enormous abilities to seeing that the artilects get built. A similar argument applies to the military and it's use of intelligent weaponry. Neither the commercial nor the military sectors will be willing to give up artilect research unless they are subjected to extreme Terran pressure.

To the Terrans, building artilects will mean taking the risk that the latter may one day decide to exterminate human beings, either deliberately or through indifference. The only certain way to avoid such a risk is not to build them in the first place. The Terrans will argue that human beings will fear the rise of increasingly intelligent machines and their alien differences. To build artilects will require an "evolutionary engineering" approach. The resulting complexities of the evolved structures that underlie the artilects will be too great for human beings to be able to predict the behaviors and attitudes of the artilects towards human beings. The Terrans will be prepared to destroy the Cosmists, even on a distant Cosmist colony, if the Cosmists go ahead with an advanced artilect building program.

In the short to middle term, say the next 50 years or so, the artificial brain based industries will flourish, providing products that are very useful and very popular with the public, such as teacher robots, conversation robots, household cleaner robots, etc. In time, the world economy will be based on such products. Any attempt to stop the development of increasingly intelligent artilects will be very difficult, because the economic and political motivation to continue building them will be very strong in certain circles. If the brain-based computer industries were to stop their research and development into artilects, then many powerful individuals, including the artilect company presidents and certain politicians will lose big money and political influence. They will not give up their status without a fight.

However, as the intelligence levels of the early artilects increases, it will become obvious to everyone that the intelligence gap between these artificial-brain-based products and human beings is narrowing. This will create a growing public anxiety. Eventually, some nasty incident or series of incidents will galvanize most of society against further increase of artificial intelligence in the artilects, leading to the establishment of a global ban on artilect research.

The Cosmists however, will oppose a ban on the development of more intelligent artilects, and will probably go underground. If the incidents continue and are negative enough, the anger and hatred of the Terrans towards the Cosmists will increase to the point where the Cosmists may decide that their fate is to leave the Earth, an option that is quite realistic with 21st century technology.

Since the Cosmists will include some of the most brilliant and economically powerful people on the planet, they will probably create an elite conspiratorial organization whose aim is to build artilects secretly.

The book presents a scenario that the author feels to be the most plausible to him, although it is highly debatable and obviously not the last word. This scenario goes as follows. The Cosmists create an asteroid-based colony, masked by some innocuous activity. In reality, this secret society devises a weapon system superior to the best on the Earth. With their wealth and the best human brains, this may be achievable. They will also start making advanced artilects. If the Terrans on the Earth discover the true intentions of the Cosmists, they will probably want to destroy them, but not dare to because of the counter threat of the Cosmists with their more advanced weapons. The stage is thus set for a major 21st century war in which billions of people die – "gigadeath."

This horrific number is derived from an extrapolation up the graph of the number of deaths in major wars from the beginning of the 19th century to the end of the 21st century. Approximately 200 million people died for political reasons -- wars, purges, genocides, etc., in the 20th century.

It is worrying about the possibility of a gigadeath "Artilect War" between the Terrans and the Cosmists, based indirectly on my own brain building work, that keeps me awake at night. Hopefully by writing this book, and perhaps also by making a movie whose central plot is based on the main ideas of this book, the global public will be made aware of what is coming, and will debate the topic hotly.

Since ultimately, I am a Cosmist, I do not want to stop my work. I think it would be a cosmic tragedy if humanity freezes evolution at the puny human level, when we could build artilects with godlike powers. However, I am not a 100% Cosmist. I shudder at the prospect of gigadeath and hence feel that the general public should be warned in time before the artilects are among us.

This book is part of that warning.

The profound schizophrenia that I feel on the Cosmist/Terran species dominance issue will be felt by millions of people within a few years I expect. There is probably Cosmist and Terran in nearly all of us, which may explain why this issue is so divisive. I am simply one of the first to feel this schizophrenia. Within a decade it may be all over the planet.

I close this last chapter of the book with a repetition of the pithy slogan introduced in the first chapter. It summarizes the two main viewpoints in the artilect debate in a nutshell, a debate that I believe will be raging in the coming decades. Here it is again --

"Do we build gods, or do we build our potential exterminators?"

Glossary

This book is full of terms new to the reader. This is because "new ideas need new labels" and this book contains many new ideas. It may be useful therefore to assemble all these new terms in a glossary so that readers can consult them as the need arises. Alternatively, the glossary may serve as an object of study in its own right. Not all of the terms listed and defined here have been coined by me. Those that are NOT original are asterisked (*).

Anthropic Principle (*)   The strong form of this principle states that the values of the constants in the laws of physics are so fantastically improbably finely tuned to allow life to exist in our universe, that it looks plausible that the universe was designed with life in mind, i.e. that it was created by some godlike being. The anthropic principle has made a lot of physicists and astronomers more tolerant of traditional religious beliefs in an omnipotent creator.

Artificial Brain (*)   My main aim in life is to make artificial brains, which I do in the following way. I evolve neural net circuit modules directly in hardware at hardware speeds in about a second and then assemble zillions of them in humanly defined artificial brain architectures to control robots. My fear that long term brain building will lead to a major war late this century has been one of the main motives for me to write this book, i.e. to sound the alarm.

233

Artificial Embryology A hypothesized technology in which engineers use techniques borrowed from the biological study of embryology, i.e. how fertilized eggs grow into animals and plants, to build products. A form of artificial DNA will provide growth instructions for artificial cells to grow, divide, and differentiate, to build complex structures. (See the definition for "embryofacture" below). Artificial embryology offers an alternative engineering approach to using zillions of nanoscale robots (nanots) to build nanotech-based products.

Artilect An abbreviation of the term "artificial intellect," or "artificial intelligence" or "massively intelligent machine." Building artilects will be made possible by 21st century technologies. I believe that a major war over the issue of whether artilects should be built or not will take place by the end of this century. This book's theme is a discussion of the likelihood of an "Artilect War," hence the book's title.

Artilect Era A hypothesized, possibly post-human, era in which the artilects have become the dominant species. This book speculates a little on what these artilects might do to occupy their immortal lives. Their godlike abilities might allow them to experiment with the creation of new universes. Perhaps our universe is the toy of an advanced artilect.

Artilect Issue The issue is whether or not human beings should build artilects this century. As 21st century technologies enable increasingly the construction of artilects, an "artilect debate" will arise. Human society will split into two major groups, the "Cosmists," in favor of building them, and the "Terrans," opposed. I believe that disagreements on this issue will become so strong and passionate, that they will very probably lead to a major war, the "Artilect War," before the end of the 21st century.

Artilect Productions Inc   I am a member of a small independent Hollywood movie company called "Artilect Productions Inc." My personal motivation in helping to make a movie on the artilect theme is to spread to a worldwide audience, the idea that an Artilect War may be coming, so that people will start thinking about it. I want to see the creation of what I call the "artilect debate." A Hollywood movie, if it can be got off the ground, would be the most effective means of spreading the word. We want to call the movie, "Artilect."

Artilect War   A hypothesized war late in the 21st century between two human groups, the "Terrans" and the "Cosmists" over the issue of whether or not humanity should build artilects. For the Cosmists, building artilects will be like a religion. To the Terrans, building artilects will be the creation of humanity's potential exterminators. The Cosmists will be inspired by awe, the Terrans by fear. Since the stake is the survival of the human species, passions will be extreme, and the war will be major. It may cost billions of human lives -- "gigadeath." Note the distinction between an "Artilect War" and a "Species Dominance War." See the definition below.

Big Picture (*)   In the context of this book, the "big picture" includes such things as the universe and its immensity, the idea that human beings could make artilects, that there are much bigger things than the trivial pursuits of human beings. Cosmists see the big picture. Terrans don't or don't want to. Terrans want human beings to remain the dominant species on the planet. Cosmists have a more cosmic perspective.

Big Things   The big things, in the Cosmist sense, are the possibilities of building artilects, building godlike, immortal, virtually omnipotent creatures with intelligence levels potentially trillions of trillions of times greater than the human level, which could explore the universe, answer the deep existential and scientific questions etc. The big things are the godlike things that would be within the grasp of the artilects, but only if the Cosmists are free to build them and are not stopped by the Terrans.

Brain Architect (BA)   A brain architect designs artificial brains. Its what I do. I evolve neural network circuit modules at electronic speeds and then assemble them into artificial brain architectures. To know how to do that I need to be a brain architect.

Brain Builder   I am a brain builder, i.e. a builder of artificial brains. Now that the new field of evolvable hardware -- see the definition below -- is established, I can evolve neural net circuit modules at electronic speeds. This is fast enough to make practical the creation of zillions of them in a reasonable time and then to put them into humanly defined artificial brain architectures. I believe that the brain builder industry will eventually dominate the world economy, as brain-based products with growing artificial intelligence become very popular with the public. The brain builders will be responsible for the creation of the artilect problem.

Brain Building Industry   Once the brain building pioneers show that the concept of building artificial brains is valid, it is very likely that industry will get into the brain building business, creating such products as household cleaner robots, teacher robots, conversational robots, etc. Brain-based computers will eventually dominate the computer industry and be worth more than a trillion dollars worldwide by the year 2020. Well into this century, the brain building industry will be the mainstay of the global economy.

Climbing Time   The amount of time in years between a) the moment when the artilects show the first signs of intelligence and

236

b) the moment when artilectual intelligence really begins to shoot up exponentially, a moment usually called "the singularity."

If the "climbing time" is short, say less than five years, that will probably not be long enough for the artilect debate to really heat up, and hence there will be no Artilect War between the Terrans and the Cosmists. If so, there may still be a "species dominance war" between human beings and the advanced artilects. The whole point, from the Terran perspective, of an Artilect War between Cosmists and Terrans, is that the Terrans might have some chance of winning it, by defeating the Cosmists. The Terrans want to avoid the risk of a species dominance war occurring between human beings and the artilects, a war that the Terrans would have zero chance of winning. At least with an "Artilect War" between Cosmists and Terrans, the human species may survive.

My opinion is that the human brain is so complex that it will take a good 10 years, and probably a lot longer, of "climbing time" for the brain builders to construct artilects of human level intelligence, once they have constructed the first "interestingly" intelligent artilects. This assumes that the brain builders will be basing their artilect designs upon neuroscience principles that need to be discovered. Copying the human brain as closely as possible is one sure way of producing human level artificial intelligence.

I think the artilect debate could start and heat up to an Artilect War within a period of 10-20 years, if the climbing time were as short as that. I also believe that the start of the climbing time will not be for several decades, say not before 2030, i.e. some 10 years after we have true nanotechnology, and which should be enough time for "interestingly" intelligent artilects, based on nanotech, to be built. By that time, space technology and space transport economics ought to have advanced enough for the Cosmists to be able to leave the planet in sufficient numbers to be an effective force against the Terrans.

Complexity Independence Complexity independence is the name I give to the idea that the internal complexity of a system that is being evolved using the "evolutionary engineering" approach is irrelevant to the evolutionary algorithm that is evolving it. This means that this internal complexity can be higher than the level that human engineers could design or understand. This greater complexity can thus create greater functionality. I believe that complexity independence is the great strength of evolutionary engineering. It may eventually dominate this century's engineering, as we build more and more complex systems, in such domains as brain building, nanotechnology, embryofacture, etc.

Cosmist A Cosmist is a person who wants to see artilects built. Cosmists are opposed to the Terrans. The Terrans will fear that artilects may one day decide that the human species is a pest and then exterminate "it." The only sure way to avoid this risk is to place a total ban on the creation of artilects beyond a certain intelligence level. Cosmists are prepared to take this risk for the sake of creating godlike artilects with intellectual capacities trillions of times greater than human beings. Cosmists conceive their ideas as a religion, which motivates them powerfully. Cosmists see the "big picture" and want to do "big things." See the definitions above.

Cosmist-Terran Dichotomy The Cosmist-Terran dichotomy is the bitter ideological dispute between the Cosmists and the Terrans this century, leading eventually to the likelihood of a major war. This ideological clash will be based on the issue of whether humanity should or should not use its 21st century technologies to build artilects. Cosmists say yes, Terrans say no.

Cosmist Transition The transition of the dominant life form on a planet from biological to artilectual. This transition may have occurred zillions of times throughout our universe, as intelligent biological species, such as ourselves, reach a level of technological sophistication whereby the construction of artilects becomes possible. Since the invention of nuclear weapons would probably closely precede the rise of artilect technology, plus the fact that artilect creation would imply the creation of a new dominant species, the previously dominant biological species concerned would probably object, and fight a "species dominance war." Perhaps many intelligent biological species do not survive these "Cosmist Transitions." (Also known as, "Species Transition," "Artilect Transition," "Artilectual Transition").

Cosmosia A suggested name, pronounced "cos-mo-sha," for the Cosmist colony, once the artilect debate has heated up enough for the Terran and the Cosmist communities to have geographically separated -- whether the colony is on the Earth, or as is more likely, in space, and probably deep space.

Creeping Cosmism This idea expresses the difficulty with which Cosmist doctrine can be stopped or even slowed. There will be so much economic, political and military momentum behind the building of artificial brains by the middle of this century, that decelerating this process will be extremely difficult, requiring a very powerful counterforce. This counterforce should appear in the form of a mass Terran fear of the artilect. Creeping cosmism will occur because nearly all institutions will want their computers to be "just a little smarter" to solve this problem or that problem. This process will continue until some kind of incident or crisis occurs to stop it.

239

**Cyborg (*)**   An abbreviation of the term "cybernetic organism," i.e. "part human, part machine," or "part natural, part artificial" People can become Cyborgs by adding artilectual brain implants to their human brains. The early Cyborgs can be seen as a third category of people to the Terrans and the Cosmists, or as a subgroup of the Cosmists. Some Cosmists may prefer to "build themselves" stepwise into artilects than to build artilects external to themselves.

There seems to be an obvious alliance to be made between Cyborgs and the Cosmists, provided that the Cyborgs are not too advanced, not too artilectual, so as to leave humans far behind in performance terms. Both Cyborgs and Cosmists want to build artilects. Both share the same dream, with the one difference that Cyborgs actually become the artilects, whereas the Cosmists remain human. Cyborgs will be as threatening to the Terrans as artilects, and will be rejected by them. Cyborgs may continue to look human, if they choose not to use genetic engineering to modify their bodies, skulls etc. However, by merely injecting a few cubic millimeters of molecular scale, 3D, heatless, one bit per atom, artilectual, brain implant, a Cyborg with a human body could become an artilect in terms of its intellectual capacities. Advanced Cyborgs and advanced artilects are effectively the same thing to the Terrans. Both could threaten humanity's survival.

**Embryofacture**   An abbreviation of the term "embryological manufacture," i.e. using artificial embryological techniques (see the definition above) to manufacture products, using nano-technological principles (see the definition below). Embryofacturing techniques will be needed if the Cosmists are to build nanotech-based artilects that self assemble.

**Embryological Engineers**   Engineers who are

240

embryofacturers. (See the above definition). Nanotech industries will need such engineers to build human, and larger, scale products that self assemble. Embryological engineers will use the same method that nature uses to build its human scale products, i.e. embryological construction.

Entropy (*) A classic term used in physics to denote the measure of disorder in a closed system, i.e. one in which no energy or matter gets in or out. The second law of thermodynamics says that in a closed system, the total entropy cannot decrease. Usually it increases, e.g. when ice melts. The fact that entropy normally increases explains the common sense phenomena that broken windows do not suddenly repair themselves, that stirred milk in coffee does not unstir itself spontaneously, etc.

Evolutionary Engineering The application of evolutionary methods to the engineering of complex systems. (See the definition for evolvability below).

Evolvability (*) The ability of systems to evolve according to an evolutionary engineer's satisfaction. Evolvability is a critical concept in the new field of evolutionary engineering. When systems are too complex for traditional top-down, blueprint-based human design methods, the only approach remaining may be that of evolutionary engineering. This approach uses evolution to build complex products and systems. If the only way to build a complex product is to use evolutionary engineering, and the evolvability of that product is low, then all is lost. Evolvability is a critical concept for an evolutionary engineer. It plays a daily role in my brain builder work when I try to evolve neural network circuit modules. Sometimes they don't evolve with the functionalities I want, so I often have to rethink, by changing the neural model I'm evolving.

Evolvable Hardware This is an idea I had in 1992, which conceives the bit string instruction that is used to configure (wire up) a programmable hardware chip as a chromosome (instruction string) of a genetic algorithm (a program that simulates the Darwinian evolution process). The fitnesses (performance qualities) of a population of programmable chips are measured and the better ones are allowed to make more copies of themselves in the next generation, while the worse ones are killed off. The "chromosomes" of the children are then randomly mutated, and the whole process loops through again. Eventually, thanks to the Darwinian survival-of-the-fittest selection pressure, functional circuits evolve. Evolvable hardware ("E-Hard" or "EH," for short) is now a thriving scientific specialty with its own conferences, academic journals, and research groups around the world. I use EH techniques to evolve the neural network circuit modules for making artificial brains.

Femtolect An abbreviation of the term "femtometer scale artilect," i.e. an artilect based on femtotech(nology) and femtometer based components. (See the definition for femtotech below).

Femtometer (*) One millionth of a billionth of a meter, the scale of quarks in nucleons (protons, neutrons, in the nucleus of the atom). Femtometer based technology might use quarks as the building blocks to make quark-gluon chemistry, perhaps in neutron stars. A femtometer is a million times smaller than a nanometer, which is the size of molecules.

Femtotech An abbreviation of the term "femtometer scale technology," i.e. of the scale of quarks inside protons, neutrons etc. Femtotechnology at the present time is only a speculation. No research work has yet been undertaken in the labs, unlike nanotechnology, which is now a thriving research area. Femtotech is probably impossible for human beings, but advanced artilects could probably achieve it. If the artilects could develop a femtotech, they could create creatures based on femtoscale phenomena (femtolects), which could outperform the nanoscale artilects (nanolects) by a factor of a trillion trillion for a given unit volume and time. The femtolect could do to the nanolect (artilect) what the nanolect (artilect) could do to human beings this century, i.e. replace them, by becoming the next dominant species.

__Globa__ Globa is the name given to the global state with its own global court to settle international disputes. Globa would have its own global armed force to police those settlements. The advance of technology forces the growth of the size of political autonomous units. The logical conclusion of this process is when the unit is of planetary size. Globa will banish major wars from the planet and will spread material affluence and happiness to the globe. The Globa concept is an example of techno-optimism, as distinct from the techno-pessimism of the Artilect War.

__Homebots__ An abbreviation of the term "home robot." Once the early artilects get smart enough and useful enough, household robots will be extremely popular and in high demand by the public. A huge homebot industry will be created to research and develop homebots. As the homebots get smarter and smarter every year, the general public will become alarmed at the rise of artilectual intelligence and wonder whether and when its increase should be stopped. As homebots and industrial robots become smarter, they will replace human workers, being more efficient, never tiring, and never complaining. Major reshuffles of the workload will follow, creating as much of a social disturbance as the industrial revolution in the 18th century. Human beings will be freed from the drudgeries of boring, dirty, dangerous work, filling their time with leisurely pursuits and receiving government handouts to live. However, there will only be a few decades in which this major social transition can take place, because once the homebots and industrial robots become smart enough to replace human workers, they will not remain at that intelligence level for very long. Once artilects become smart, they will soon become very smart within a few decades or less.

One might speculate whether artilects themselves might use homebots and industrial robots for their own convenience. Presumably in any society, whether human or artilectual, there would be a need for robots of different levels of intelligence possessing different skill levels.

Intelligence Theory A hypothesized theory of the nature of intelligence. Once neuroscience understands why intelligence levels of human beings differ because of differences in their neural structures, it will be possible to create an intelligence theory, which can be used by neuro-engineers to increase the intelligence of the artificial brains they build.

Mono A mono-lingual, mono-cultured person, who has lived in only one culture or country. Monos are limited as individuals by the limitations of the monoculture that programs them. A contrast can be made between "monos" and "multies." A multi is a multi-lingual multi-cultured person.

Moore Doublings A consequence of Moore's Law (see next definition) is that after many doublings of electronic performance levels, e.g. density of components on a chip, the speed of chips, etc., the absolute size of each doubling soon becomes enormous. For example, if one doubles and doubles the number 2, after 20 doublings, the figure is over a million, after 40 doublings, the figure is over a trillion. The growth in electronics potential is thus not a linear one but is rather exponentially explosive. This growth will enable the technology for building artilects this century.

Moore's Law (*) Gordon Moore was one of the cofounders of the Intel microprocessor corporation who noticed in the mid 1960s that the performance of integrated circuits (ICs) was doubling roughly every 18 months or so, due to a down-scaling of the size of electronic components. This trend has remained roughly true for the past 40 years and fuels the economic growth of our times. If Moore's Law continues until 2020, we will be able to store a bit of information on a single atom. Moore's Law will enable the building of artilects this century, and hence initiate the artilect debate and possibly the Artilect War.

Multi A multi-lingual, multi-cultured person, who has lived in several cultures or countries. Multies are enriched by the strengths of several cultures they absorb into their personalities. Multies usually prefer the company of other multies, taking an attitude that "Monos are boring!" Monos tend to have a poor sense of cultural relativity, unconsciously imposing the mono-cultured standards of their mono-cultured upbringing upon the values and behaviors of the multies. This failing of the monos is seen as unsophisticated and limiting by multies.

Nanolect An abbreviation of the term "nanoscale artilect," i.e. one based on nanotechnological components and principles, as distinct from a femtolect, which is based on femtotech and femtometer based components. A nanolect would be as inferior to a femtolect as would a human being to a nanolect. In this book, an artilect is usually conceived of as a nanolect, not a femtolect. Femtotechnology has barely been speculated upon by the scientists let alone explored in the laboratories. A nanolect could be of

245

virtually any size, ranging from the submicroscopic, requiring a microscope to see it, to asteroid size (hundreds of kilometers across) and beyond. The more atoms used to store information and artilect circuitry, the larger the size of the nanolect. Atoms have a definite size and take up space.

Nanometer (*) Nano means a billionth. A nanometer is a billionth of a meter, i.e. molecular scale. Atoms are about a tenth of a nanometer across. Today's computers function at nanosecond speeds.

Nanotech (*) An abbreviation of the term nanotechnology, or "molecular scale engineering," which is the construction of molecular scale machines, one atom at a time, with perfect atomic precision. Artilects will be based on nanotech.

Nanots An abbreviation of the term "nanoscale robot," i.e. a robot of molecular scale (nanometer, or billionth of a meter), capable of picking up single atoms and placing them with atomic precision to build molecular components and machines.

Netilect An abbreviation of the term "network of artilects." Artilects, even if they can store one bit on one atom, will face computational limits on how much data they can store within a given mass. They will be able to exchange data, experience, and ideas with each other by linking up in a huge network, probably via the use of electromagnetic waves, or some physical phenomenon human beings have not discovered yet, or may never discover due to human intellectual limitations.

Phys-Comp (*) An abbreviation of the term "physics of computation," which investigates the fundamental physical limits of computation, e.g. "Is it possible to compute with zero heat generation?" "What is the maximum rate of computation one can perform in a given volume, in a given time?" etc. Quantum computing questions are much discussed in the field of phys-comp.

Quantum Computing (*) As the electronics industry shrinks its components to molecular scale, quantum phenomena begin inevitably to appear. Quantum computing is a form of computation that takes advantage of the quantum phenomenon of superposition of quantum states, which can handle many classical mechanical cases at once. This is hugely more efficient than classical computing, which computes one case at a time. Nanotech based artilects will need to be quantum computers, given their molecular scale.

Reversible Computing (*) A style of computing that generates zero heat, and hence serves as the basis for 3D-computer circuitry. Reversible computing employs reversible logic gates that do not destroy information. Thermodynamical considerations show that wiping out bits of information generates heat. By sending information through a reversible computer, copying the answer, and then sending the result back (reversibly) through the same circuitry, one arrives at the original input. No information is destroyed, so no heat is generated. Reversible computing is inevitable because molecular scale circuits will explode if they employ traditional irreversible techniques. This century's artilects will be based upon this computing style.

Scaling War The type of war that may occur when new technologies allow the construction of new life forms of vastly superior intellectual, and other capacities. The greater intellectual capacities result from greater signaling speeds and component densities. As an example of a scaling war, take the "Artilect War" between the Cosmists and the Terrans, both human groups, when nanotech-based artilects (nanolects) become possible. A further example could be a war between the Xists and the Yists (both

247

nanolect groups) when femtotech based artilects (femtolects) become possible. The basic cause of a scaling war is due to disagreements between two groups of the same scale, over the risks that may follow if superior creatures, based on the new (smaller, faster, denser) technologies, are built. The latter might decide to exterminate the creatures based on the inferior (larger, slower, sparser) technologies.

Scientist's Religion Most scientists are not conventionally religious, tending to look upon conventional religions as being based on prescientific superstitions that are incompatible with both modern scientific knowledge and with the modern scientific approach of testing hypotheses. However, scientists are also human beings, and hence have the same religious hungers as other people. Cosmism is a religion to the Cosmists, yet is based on modern science. It could be a scientist's religion in the sense of providing a sense of awe, of collective purpose, the creation of gods, etc. but be consistent with science. Cosmism is a set of "religious" beliefs that scientists could find credible.

Singularity (*) In mathematics, a singularity is a value that approaches infinity. In the context of the artilect debate, the singularity refers to the idea that a time will come when a machine is made which is so smart that it will be able to redesign itself better and faster than human beings. The result will be a superior machine that then designs an even better one, ad infinitum, and at the speed of light. Another variant is that the machine will be an excellent learner, and simply starts learning for itself at a rate a million times faster than humans. Its increase in intelligence and knowledge would seem infinitely fast to humans.

Smartilect An abbreviation for a "smart artilect." There will probably be many kinds of artilect built this century. The early

248

artilects will be less intelligent and capable than the later ones. The really smart ones could be called "smartilects."

__Space Consciousness__ Space consciousness is the emotionally overwhelming feeling one has when one sees either through vivid, large scale, 3D graphics, or for real to some extent, the immensity of space, the billions of stars in a galaxy, the billions of galaxies in our universe. This feeling makes one conscious of the total insignificance of human preoccupations. Space consciousness makes human beings more aware that there are "bigger things" in life and in the universe than normal day-to-day human activities and goals. Space consciousness is an important concept to the Cosmists and to Cosmist ideology.

__Species Dominance Debate__ The issue that will dominate global politics this century is that of species dominance. The species dominance debate will focus upon the question of whether humanity should build artilects or not. Should human beings remain the dominant species, or should artilects be built to surpass us in intelligence levels. Progress in 21st century technologies will allow artilects to be built, and force the dominant species issue to be debated.

__Species Dominance War__ A war between the dominant biological species of a planet and the artilects that it creates. Since the artilects would be a superior species to the biologically based creatures, the war would settle which species would dominate. A distinction needs to be made between a "Species Dominance War" and the "Artilect War." The "Artilect War" is the name given to the war between two human groups fighting over whether or not artilects should be built. A "species dominance war" could occur anywhere in the universe and is a far more general concept. For example, if "femtolects," i.e. "femtotechnology" based "artilects," had a war with the "nanolects," i.e. "nanotechnology" based "artilects," that would be a "species dominance war."

Talkies A talkie is a slang word for a conversational robot, i.e. one that a human being can chat with, have a friendship with, and if the talkies are smart enough, to have a relationship with. Talkies will need to be of almost human level intelligence to be really effective and popular with their human owners.

Teacherbots A teacherbot is a teacher robot, i.e. an artificially intelligent robot capable of teaching human beings, by adapting to their intelligence levels, interests, and motivations.

Terran A Terran, based on the word "terra," the Earth, is someone who feels that artilects are too potentially dangerous to the survival of human beings to be built. Terrans oppose the Cosmists, who feel that they should be built. The Terrans want human beings to remain dominant species on the Earth. The fundamental attitude of the Terrans is that the only way to be sure that there is zero risk that the artilects will destroy the human species at some later date, is that the artilects are never built in the first place. Therefore to preserve the survival of the human species, the Terrans will stop the Cosmists, no matter what the cost, even if a major war is necessary.

Terran Problem The Terran problem is the name given to the fear of the leaders of the artificial brain based industries that they may see their companies lose sales and political influence if the Terrans succeed in placing a global ban on the development of artilects beyond a certain "safe-for-humans" intelligence level.

Bibliography

Books on Cyborgs

Andy CLARK, *Natural-Born Cyborgs: Minds, Technologies, and the Future of Human Intelligence,* Oxford University Press, 2003.
Chris H. GRAY, *Cyborg Citizen,* Routledge, 2001.
Jacques HOUIS et al (eds.), *Being Human: The Technological Extensions of the Body,* Marsilio, 1999.
Kevin WARWICK, *I, Cyborg,* Century, 2002.

Books on Intelligent Machines, Posthumanity, etc.

Azamat ABDOULLAEV, *Artificial Superintelligence,* EIS Ltd., 1999.
Igor ALEXANDER, *How to Build a Mind,* Weidenfeld and Nicholson, 2000.
James BAILEY, *After Thought: The Computer Challenge to Human Intelligence,* Basic Books, 1996.
Peter BOCK, *The Emergence of Artificial Cognition,* World Scientific, 1993.
Damien BRODERICK, *The Spike: How Our Lives are Being Transformed by Rapidly Advancing Technologies,* Forge, 2001.
Rodney A. BROOKS, *Flesh and Machines: How Robots Will Change Us,* Pantheon, 2002.
Maureen CAUDILL, *In Our Own Image: Building an Artificial Person,* Oxford University Press, 1992.

251

Thomas M. GEORGES, *Digital Soul: Intelligent Machines and Human Values,* Westview, 2003.
Jerome C. GLENN, *Future Mind: Artificial Intelligence,* Acropolis, 1989.
Ray KURZWEIL, *The Age of Intelligent Machines,* MIT Press, 1990.
Ray KURZWEIL, *The Age of Spiritual Machines: When Computers Exceed Human Intelligence,* Viking, 1999.
Ray KURZWEIL, *Are We Spiritual Machines? Ray Kurzweil vs. the Critics of Strong A.I.,* Discovery Institute, 2002.
James MARTIN, *After the Internet: Alien Intelligence,* Capital Press, 2000.
Pamela McCORDUCK, *Machines Who Think,* Freeman, 1979.
Bill McKIBBEN, *Enough: Staying Human in an Engineered Age,* Timers Books, 2003.
Hans MORAVEC, *Mind Children: The Future of Robot and Human Intelligence,* Harvard University Press, 1990.
Hans MORAVEC, *Robot: Mere Machine to Transcendent Mind,* Oxford University Press, 1999.
Douglas MULHALL, *Our Molecular Future: How Nanotechnology, Robotics, Genetics, and Artificial Intelligence Will Transform Our World,* Prometheus Books, 2002.
Gregory S. PAUL, Earl D. COX, *Beyond Humanity CyberEvolution and Future Minds,* Charles River Media, 1996.
Sidney PERKOWITZ, *Digital People: From Bionic Humans to Androids,* Joseph Henry Press, 2004.
Peter RUSSELL, *The Global Brain Awakens: Our Next Evolutionary Leap,* Element, 2000.
Jeffrey SATINOVER, *The Quantum Brain: The Search for Freedom and the Next Generation of Man,* Wiley, 2001.

Geoff SIMONS, *Is Man a Robot,* Wiley, 1986.
Gregory STOCK, *Metaman: The Merging of Humans and Machines into a Global Superorganism,* Simon & Schuster, 1993.
John TAYLOR, *The Shape of Minds to Come,* Michael Joseph, 1971.
Kevin WARWICK, *March of the Machines: Why the New Race of Robots will Rule the World,* Century, 1997.
Kevin WARWICK, *In the Mind of the Machine: The Breakthrough in Artificial Intelligence,* Arrow, 1998.

Books on Nanotechnology

William I. ATKINSON, *Nanocosm: Nanotechnology and the Big Changes Coming from the Inconceivably Small,* Amacom, 2003.
B. C. CRANDALL et al., *Nanotechnology: Research and Perspectives,* MIT Press, 1992.
B. C. CRANDALL (ed.), *Nanotechnology: Molecular Speculations on Global Abundance,* MIT Press, 1996.
K. Eric DREXLER, *Engines of Creation: The Coming Era of Nanotechnology,* Anchor Press, Doubleday, 1986.
K. Eric DREXLER et al., *Unbounding the Future: The Nanotechnology Revolution,* Morrow, 1991.
K. Eric DREXLER, *Nanosystems: Molecular Machinery, Manufacturing, and Computation,* Wiley Interscience, 1992.
Sandy FRITZ, *Understanding Nanotechnology,* Warner Books, 2002.
Michael GROSS, *Travels to the Nanoworld: Miniature Machinery in Nature and Technology,* Plenum Trade, 1999.

Markus KRUMMENACKER et al (eds.) *Prospects in Nanotechnology: Toward Molecular Manufacturing,* Wiley, 1995.

Wil McCARTHY, *Hacking Matter: Levitating Chairs, Quantum Mirages, and the Infinite Weirdness of Programmable Atoms,* Basic books, 2003.

Mark RATNER et al. *Nanotechnology: A Gentle Introduction to the Next Big Idea,* Prentice Hall, 2003.

Ed REGIS, *Nano: The Emerging Science of Nanotechnology, Remaking the World – Molecule by Molecule,* Little,Brown, 1995.

Edward RIETMAN, *Molecular Engineering of Nanosystems,* Springer, 2001.

Gregory TIMP (ed.), *Nanotechnology,* Springer, 1999.

Books on Quantum Computing

Amir D. ACZEL, *Entanglement: The Greatest Mystery in Physics,* Four Walls Eight Windows, 2001.

Julian BROWN, *The Quest for the Quantum Computer,* Touchstone, 2000.

Mika HIRVENSALO, *Quantum Computing,* Springer, 2001.

Gerald J. MILBURN, *Schrodinger's Machines,* Freeman, 1997.

Gerald J. MILBURN, *The Feynman Processor: Quantum Entanglement and the Computing Revolution,* Perseus Books, 1998.

Michael A. NIELSEN and Isaac L. CHUANG, *Quantum Computation and Quantum Information,* Cambridge University Press, 2000.